Metric and other conversion tables

G. E. D. Lewis, B.Sc., Ph.D., F.R.G.S.

Longman

LONGMAN GROUP LIMITED
London

Associated companies, branches and representatives throughout the world

© Longman Group Limited 1973

First published 1973

ISBN 0 582 42940 4

Library of Congress Catalog Card Number 73–86518

Set in Univers
and printed in Great Britain
by William Clowes & Sons Limited London, Colchester and Beccles

CONTENTS

INTRODUCTION

The metric system of weights and measures is based on the metre as the unit of length and was first adopted in France during the French Revolution. Unlike the British imperial system, which is complex with many different units, the metric system is simple because it is decimal throughout—all units being obtained by multiplying by factors of ten. For example, among everyday units:

    a *milli*metre is *one thousandth of* a metre,
    a *centi*metre is *one hundredth of* a metre,
    a *kilo*metre is *one thousand* metres,
    a *hectare* is *one hundred* are (ten thousand square metres),
    a cubic centimetre is one millionth of a cubic metre,
    a *milli*litre is *one thousandth of* a litre,
    a *centi*litre is *one hundredth of* a litre,
    a *hecto*litre is *one hundred* litres,
    a gram is one thousandth of a kilogram,
    a tonne is one thousand kilograms,
    a *kilo*watt is *one thousand* watts.

Most countries now use or have agreed to adopt a common system of metric units. This is a modern form of the metric system and is known as 'le Système International d'Unités' (International System of Units) or SI, which was the name given in 1960 by an international body responsible for maintaining standards of measurements. It is anticipated that eventually the whole world will adopt the metric system of SI units.

Although most people will soon be thinking in metric units, the need to convert from the old system to the new, and vice versa, will remain for many years. There is therefore, at the present time, a special need for a book of conversions. This book has been compiled with this object in mind especially for the use of students at all levels, particularly in technical and practical subjects, and for normal commercial use. It is also designed for geographers, estate agents and a wide range of engineers.

It contains tables which are frequently used and is so designed that conversions can be easily made from British (or American where they differ from the British) units to SI units and vice versa. Unless they are exact terminating numbers, the list of conversion factors on pages 5 and 6 have been expressed to the nearest sixth significant figure. A more accurate conversion factor used in the computation of a particular table is given at the head of the table. The values in the tables, unless they are exact, have been rounded to the number of decimal places tabulated. The number of significant figures, and the range of magnitude of the tables are determined by their practicability. For example, Tables 21 and 22 cover a wide range because on Ordnance Survey plans, areas of land are given in hectares to three places of decimals; similarly Tables 91 and 92 cover the recorded extremes of atmospheric pressure.

The conversion factors, definitions and symbols or abbreviations are based mainly upon 'Changing to the Metric System' (Anderton and Bigg, N.P.L.) (1972 H.M.S.O.). All the tables have been independently computed and great care has been taken to ensure their accuracy. They have also, where possible, been checked against reliable authorities such as 'BS 350, Conversion Factors and Tables', Part 1: 1959, Part 2: 1962, and Supplement No. 1 (1967) published by the British Standards Institution.

The publishers regret that they cannot hold themselves responsible for any inconvenience or loss occasioned by any errors or omissions. The author would, however, be grateful if readers informed him of any errors, and gratefully acknowledges the invaluable assistance already received from E. H. Varley, M.Sc., and Mrs. P. Anderton of the National Physical Laboratory.

THE INTERNATIONAL SYSTEM OF UNITS (SI)

The International System of Units, abbreviated SI, includes (i) primary or base units, (ii) supplementary units, (iii) derived units, and (iv) decimal multiples and sub-multiples of these units.

(i) The SI is based on six primary units:

Quantity	Name of unit	Symbol
length	metre	m
mass	kilogram	kg
time	second	s
electric current	ampere	A
thermodynamic temperature	kelvin	K
luminous intensity	candela	cd

(ii) The SI units for plane angle and solid angle, the radian (rad) and steradian (sr), respectively, are called supplementary units.

(iii) All other SI units are derived from the six basic units and are expressed in terms of these basic units, for example, the SI unit for volume is the cubic metre (m^3). Some of these derived units have special names and symbols, for example, the newton (N) is the SI unit of force. The following are the ones most likely to be met:

Quantity	SI unit	Symbol	Expressed in terms of SI base units or derived units
area	square metre	m^2	
volume	cubic metre	m^3	
velocity	metre per second	$m\ s^{-1}$	
density	kilogram per cubic metre	$kg\ m^{-3}$	
acceleration	metre per second squared	$m\ s^{-2}$	
frequency	hertz	Hz	$1\ Hz = 1\ s^{-1}$
force	newton	N	$1\ N = 1\ kg\ m\ s^{-2}$
pressure and stress	pascal	Pa	$1\ Pa = N\ m^{-2}$
energy, work, quantity of heat	joule	J	$1\ J = 1\ N\ m$ $= 1\ kg\ m^2\ s^{-2}$
power	watt	W	$1\ W = 1\ J\ s^{-1}$ $= 1\ kg\ m^2\ s^{-3}$
electric charge	coulomb	C	$1\ C = 1\ A\ s$
electric potential difference	volt	V	$1\ V = 1\ W\ A^{-1}$
electric capacitance	farad	F	$1\ F = 1\ A\ s\ V^{-1}$
electric resistance	ohm	Ω	$1\ \Omega = 1\ V\ A^{-1}$

Note:
(a) The degree Celsius ($^\circ$C) is the customary unit for temperature and is the same unit as the degree Centigrade.
(b) The name 'degree Kelvin' ($^\circ$K) was changed to 'kelvin' (K) in 1967.
(c) The minute, the hour, the day, etc. are not SI units but are internationally recognised and therefore may be used.
(d) Similarly the common units of the angular degree (degree, minute, second etc.) will remain in use.

(e) Other multiples and sub-multiples of SI units with special names which are likely to persist and are acceptable non-SI units, include the following:

Quantity	SI unit	Symbol	Definition of unit
area	hectare	ha	10^4 m^2
volume	litre	l	1 dm^3 = 10^{-3} m^3
	millilitre	ml	1 cm^3 = 10^{-6} m^3
mass	tonne	t	10^3 kg = Mg
pressure	bar	bar	10^5 N m^{-2}
	millibar	mbar	10^2 N m^{-2}
energy, work	kilowatt hour	kW h	3.6 MJ

(iv) Prefixes are used to indicate multiples of units in powers of ten. Fourteen such prefixes have been recommended. The ones most likely to be met are:

Multiple		Prefix	Symbol	Example
1 000 000 000	= 10^9	giga	G	gigawatt (GW)
1 000 000	= 10^6	mega	M	megawatt (MW)
1 000	= 10^3	kilo	k	kilogram (kg)
100	= 10^2	hecto	h	hectare (ha)
0.1	= 10^{-1}	deci	d	decimetre (dm)
0.01	= 10^{-2}	centi	c	centimetre (cm)
0.001	= 10^{-3}	milli	m	millibar (mbar)
0.000 001	= 10^{-6}	micro	μ	microsecond (μs)
0.000 000 001	= 10^{-9}	nano	n	nanometre (nm)
0.000 000 000 001	= 10^{-12}	pico	p	picofarad (pF)

Notes on the Use of Units and Symbols

1. Names of units are not written with a capital letter, even when named after a person, e.g. 100 watts 5 amperes.
2. The symbol for a unit named after a person has a capital initial letter; e.g. N for newton, named after Isaac Newton; W for watt named after James Watt.
3. Symbols for units do not have a plural form with added 's'; e.g. 56 millimetres is written 56 mm and the full stop is not used except at the end of a sentence.
4. One space is placed between a number and its associated symbol; e.g. 99 kg.
5. For the expression of symbols for derived units, an index notation is recommended; e.g. a velocity of 5 metres per second would be written 5 m s^{-1} rather than 5 m/s.
6. The comma is not used to separate groups of three digits but they may be separated by spaces; e.g. speed of light = 2.997 925 $\times$ 10^8 m s^{-1}. (Note: this practice is not observed in the tables in this book owing to considerations of space).
7. The spellings 'gramme' and 'kilogramme' have recently been officially replaced by 'gram' and 'kilogram'.
8. In 1971 the pascal (Pa) was internationally recognised as the name for the newton per square metre, the SI unit of pressure.

DEFINITIONS

ampere (A)
SI unit of electric current. The ampere is the constant current which, if maintained in two straight parallel conductors of infinite length, of negligible circular cross-section, and placed at a distance of 1 metre apart in a vacuum, will produce between them a force equal to 2×10^{-7} newton per metre length.

calorie (cal)
Unit of heat, now replaced by the joule.

candela (cd)
SI unit of luminous intensity. The candela is the luminous intensity, in the perpendicular direction, of a surface of 1/600 000 square metre of a black body at the temperature of freezing platinum under a pressure of 101 325 newtons per square metre.

coulomb (C)
SI unit of electric charge. The coulomb is the quantity of electricity transported in 1 second by 1 ampere.

degree Celsius (°C)
Customary unit of temperature. The zero of this scale is the temperature of the ice point (273.15°K). The units of Celsius and kelvin temperature interval are identical.

gallon
Imperial or UK gallon is the space occupied by 10 pounds weight of distilled water of density 0.998 859 gram per millilitre weighed in air of density 0.001 217 gram per millilitre against weights of density 8.136 grams per millilitre. The US gallon is equal in volume to 231 cubic inches and is used for the measurement of liquids only.

hertz (Hz)
SI unit of frequency. The number of repetitions of a regular occurrence in 1 second.

joule (J)
SI unit of energy, including work and quantity of heat. The work done when the point of application of a force of 1 newton is displaced through a distance of 1 metre in the direction of the force (=N m).

kelvin (K)
SI unit of thermodynamic temperature. The kelvin is the fraction 1/273.16 of the thermodynamic temperature of the triple point of water.

kilogram (kg)
SI unit of mass. The kilogram is equal to the mass of the international prototype kilogram.

knot
UK knot = 1 UK nautical mile per hour.
International knot (kn) = 1 international nautical mile per hour.

litre* (l)
A name for the cubic decimetre.

metre (m)
SI unit of length. The metre is the length equal to 1 650 763.73 wavelengths in vacuum of the radiation corresponding to the transition between the energy levels $2p_{10}$ and $5d_5$ of the krypton-86 atom.

millibar (mbar)
Used in meteorology for barometric measurements and represents a pressure of 1000 dynes per square centimetre or 100 newtons per square metre ($=100$ N m^{-2}).

nautical mile
UK nautical mile = 6080 feet (=1.853 18 km). It is roughly 0.06% greater than the international nautical mile (n mile), (=1.852 km exactly), which is accepted by all countries, including the USA and the UK (1970). (Note—to avoid confusion, it is recommended that 'mile' should not be abbreviated.)

newton (N)
SI unit of force. That force which, when applied to a mass of 1 kilogram, gives it an acceleration of 1 metre per second per second ($=$kg m s^{-2}).

ohm (Ω)
SI unit of electric resistance. The resistance between two points of a conductor when a constant difference of potential of 1 volt, applied between these points, produces in this conductor a current of 1 ampere, the conductor not being the source of any electromotive force.

pascal (Pa)
SI unit of pressure. The pressure produced by a force of 1 newton applied, uniformly distributed, over an area of 1 square metre ($=$ N m^{-2}).

radian (rad)
SI unit of plane angle. The angle between two radii of a circle which cut off on the circumference an arc equal in length to the radius.

second (s)
SI unit of time. The second is the duration of 9 192 631 770 periods of the radiation corresponding to the transition between the two hyperfine levels of the ground state of the caesium-133 atom.

ton
Long ton refers to the UK ton of 2240 lb.
Short ton refers to the US ton of 2000 lb.

tonne (t)
Often referred to as the 'metric ton' in UK and USA (=1000 kg).

volt (V)
SI unit of electric potential ($=$W A^{-1}). The difference of electric potential between two points of a conducting wire carrying a constant current of 1 ampere, when the power dissipated between these two points is equal to 1 watt.

watt (W)
SI unit of power. The watt is equal to 1 joule per second (1 J s^{-1}).

* See page 42.

CONVERSION FACTORS

Length

1 inch (in)	= **25.4** millimetres (mm)
1 foot (ft)	= **0.304 8** metre (m)
1 yard (yd)	= **0.914 4** metre (m)
1 mile	= **1.609 344** kilometres (km)
1 fathom	= **1.828 8** metres (m)
1 UK nautical mile	= 1.853 18 kilometres (km)
1 international nautical mile (n mile)	= **1.852** kilometres (km)

Area

1 square inch (in²)	= **6.451 6** square centimetres (cm²)
1 square foot (ft²)	= **0.092 903 04** square metre (m²)
1 square yard (yd²)	= **0.836 127 36** square metre (m²)
1 acre	= 0.404 686 hectare (ha)
1 square mile	= 2.589 99 square kilometres (km²)

Volume and capacity

1 cubic inch (in³)	= **16.387 064** cubic centimetres (cm³)
1 cubic foot (ft³)	= 0.028 316 8 cubic metre (m³)
1 cubic yard (yd³)	= 0.764 555 cubic metre (m³)
1 UK gallon (UKgal)	= 4.546 092 litres (l)
	= 1.200 95 US gallons (USgal)
1 US gallon (USgal)	= 3.785 41 litres (l)
	= 0.832 674 UK gallon (UKgal)
1 UK bushel	= 0.363 687 hectolitre (hl)
	= 1.032 06 US bushels
1 US bushel	= 0.352 391 hectolitre (hl)
	= 0.968 939 UK bushel

Yield

1 UK bushel per acre	= 0.898 691 hectolitre per hectare (hl ha⁻¹)
1 US bushel per acre	= 0.870 776 hectolitre per hectare (hl ha⁻¹)

Angle

90 degrees (°)	= $\pi/2$ radians (rad)
	= 1.570 80 radians (rad)
1 degree (°)	= 0.017 453 3 radian (rad)

Velocity

1 foot per second (ft s⁻¹)	= **0.304 8** metre per second (m s⁻¹)
1 mile per hour (mile h⁻¹)	= **1.609 344** kilometres per hour (km h⁻¹)
1 UK knot	= 1.000 64 international knots (kn)
	= 1.151 52 miles per hour (mile h⁻¹)

Mass

1 ounce (avoir) (oz)	= 28.349 5 grams (g)
1 pound (lb)	= **0.453 592 37** kilogram (kg)
1 UK ton (long ton)	= 1.016 05 tonnes (metric tons) (t)
1 US ton (short ton)	= 0.907 185 tonne (metric ton) (t)
1 UK ton (long ton)	= **1.12** US tons (short tons)

Mass per unit area

1 pound per square inch (lb in⁻²)	= 0.070 307 0 kilogram per square centimetre (kg cm⁻²)

Specific volume

1 cubic foot per UK ton (ft³ ton⁻¹)	= 0.027 869 6 cubic metre per tonne (m³ t⁻¹)
1 cubic yard per UK ton (yd³ ton⁻¹)	= 0.752 480 cubic metre per tonne (m³ t⁻¹)
1 UK gallon per UK ton (UKgal ton⁻¹)	= 4.474 2 litres per tonne (l t⁻¹)

Volume rate of flow
1 cubic foot per minute (ft^3 min^{-1}) = **0.471 947 443 2** cubic decimetre per second (dm^3 s^{-1})
1 UK gallon per hour (UKgal h^{-1}) = 0.004 546 09 cubic metre per hour (m^3 h^{-1})

Density
1 pound per cubic foot (lb ft^{-3}) = 16.018 5 kilograms per cubic metre (kg m^{-3})
1 pound per UK gallon (lb UKgal^{-1}) = 0.099 776 3 kilogram per litre (kg l^{-1})

Force*
1 pound-force (lbf) = 4.448 22 newtons (N)
1 newton (N) = 0.224 809 pound-force (lbf)
1 kilogram-force** (kgf) = **9.806 65** newtons (N)

Pressure† stress
1 UK ton-force per square inch
   (tonf in^{-2}) = 15.444 3 meganewtons per square metre (MN m^{-2})
1 pound-force per square inch
   (lbf in^{-2}) = 6 894.76 newtons per square metre (N m^{-2})
  = 68.947 6 millibars (mbar)
1 inch of mercury (inHg) = 33.863 9 millibars (mbar)
1 millimetre of mercury (mmHg) = 2.784 50 pounds-force per square foot (lbf ft^{-2})
  = **133.322 387 415** newtons per square metre (N m^{-2})

Energy (work, heat)
1 therm (=100 000 Btu) = 29.307 1 kilowatt hours (kW h)
1 kilowatt hour (kW h) = **3.6** megajoules (MJ)
1 foot pound-force (ft lbf) = 1.355 82 joules (J)
1 joule (J) = 0.737 562 foot pound-force (ft lbf)
  = 0.000 947 817 British thermal unit (Btu)
1 British thermal unit (Btu) = 1 055.06 joules (J)
  = 251.996 calories (cal)
1 calorie (cal)†† = **4.186 8** joules (J)
  = 0.003 968 32 British thermal unit (Btu)

Power
1 horsepower (hp) = 745.700 watts (W)
1 watt (W) = 0.737 562 foot pound-force per second (ft lbf s^{-1})
  = 0.001 341 02 horsepower (hp)
  = 3.412 14 British thermal units per hour (Btu h^{-1})

Temperature
degree Fahrenheit (°F) = 9/5 degrees Celsius + 32
degrees Celsius (Centigrade) (°C) = 5/9 (°F − 32)
  = kelvin (K) − 273.15

Specific energy
1 British thermal unit per pound
   (Btu lb^{-1}) = **2.326** kilojoules per kilogram (kJ kg^{-1})

Calorific value
1 British thermal unit per cubic foot
   (Btu ft^{-3}) = 37.258 9 kilojoules per cubic metre (kJ m^{-3})

Intensity of heat flow rate
1 British thermal unit per square foot
   hour (Btu ft^{-2} h^{-1}) = 11.356 5 kilojoules per square metre hour (kJ m^{-2} h^{-1})

Conversion factors given in bold figures indicate exact value; other values are expressed to the nearest sixth significant figure.
* *Force* can conveniently be expressed in terms of the local gravitational pull on a known mass. Since gravity varies over the earth's surface, the weight of a given mass also varies. Thus the lbf and kgf are defined as the forces due to 'standard gravity' acting on bodies of mass 1 lb and 1 kg respectively, standard gravitational acceleration being taken as 32.174 ft s^{-1} or 9.806 65 m s^{-1}.
** *Kilogram-force* is called **kilopond** (**kp**) in Germany and some other European countries.
† It is common practice in the UK to refer to pressure in p.s.i., that is 'pounds per square inch'; this is a colloquial expression for 'pounds-force per square inch'.
†† This calorie is the international table calorie.

HOW TO USE THE TABLES

The Tables in this book are conventional in design and are easy to read. The use of these tables is illustrated by the examples below. *See also the footnotes at the end of the various tables.*

1. Direct conversion

Example: A spot height is marked 276 feet; what is the height in metres?
Read Table 3(b) on page 14:

276 feet = **84.124 8 metres**

2. (a) Conversion involving the use of an auxiliary table

Example: An island has an area of 211 square miles: what is this in square kilometres?
From Table 23(a) on page 37:

200 square miles = 517.998 square kilometres

Table 23(b) on page 37:

11 square miles = 28.490 square kilometres
Hence   211 square miles = **546.488** square kilometres

(b) Conversion involving the use of two auxiliary tables

Example: The area of a farm is 298.263 acres; how many hectares is this?
From Table 21(a) on page 31:

200 acres = 80.937 1 hectares

Table 21(b) on page 31:

98 acres = 39.659 2 hectares

Table 21(c) on page 32:

0.263 acres =   0.106 4 hectares
Hence   298.263 acres = **120.702 7** hectares
= 120.703 hectares (rounded off).

3 Conversion involving simple arithmetic

Example: A wind velocity of 23.8 feet per second is to be converted to miles per hour.
From Table 47(b) on page 50:

23 feet per second = 15.681 8 miles per hour
8 feet per second = 5.454 5 miles per hour
thus   0.8 feet per second = 0.545 5 miles per hour
Hence   23.8 feet per second = **16.227 3** miles per hour

4. Indirect use of a table

Example: The population density of the United Kingdom is 592 persons per square mile. Express this density in terms of persons per square kilometre.
There is no specific table to convert data of population density. However conversions of persons per square mile to persons per square kilometre and square kilometres to square miles are identical.
Thus, from Table 24(a) on page 37:

500 square kilometres = 193.051 1 square miles
and from Table 24(b) on page 37:

92 square kilometres = 35.521 4 square miles
Hence   592 square kilometres = **228.572 5** square miles
and so 592 persons per square mile = 229 persons per square kilometre.

Note re Accuracy of Values in Tables

Where the values in a table are not exact (i.e. there is no statement at the head of the table stating that the values are exact), they have been rounded off. They are therefore subject to a possible error in the last digit shown. In such cases a greater degree of accuracy can be obtained by using the conversion factor at the head of the table.

ADDITIONAL USES OF THE TABLES

Many of the tables in this book have additional uses as shown below. Thus, *Table 75, UK gallons per hour to cubic metres per hour*, can also be used for converting UK gallons to cubic metres, for example, 323 UK gallons = 1.468 39 cubic metres.

Table 1 INCHES TO MILLIMETRES

[in to mm]   1 in = 25.4 mm (exactly)
All values in Tables 1(a) and 1(b) are exact.

(a)

inches	0	100	200	300	400	500	600	700	800	900
0	—	2540	5080	7620	10160	12700	15240	17780	20320	22860
1000	25400	27940	30480	33020	35560	38100	40640	43180	45720	48260

(b)

inches	0.0	0.1	0.2	0.3	0.4	0.5	0.6	0.7	0.8	0.9
0	—	2.54	5.08	7.62	10.16	12.70	15.24	17.78	20.32	22.86
1	25.40	27.94	30.48	33.02	35.56	38.10	40.64	43.18	45.72	48.26
2	50.80	53.34	55.88	58.42	60.96	63.50	66.04	68.58	71.12	73.66
3	76.20	78.74	81.28	83.82	86.36	88.90	91.44	93.98	96.52	99.06
4	101.60	104.14	106.68	109.22	111.76	114.30	116.84	119.38	121.92	124.46
5	127.00	129.54	132.08	134.62	137.16	139.70	142.24	144.78	147.32	149.86
6	152.40	154.94	157.48	160.02	162.56	165.10	167.64	170.18	172.72	175.26
7	177.80	180.34	182.88	185.42	187.96	190.50	193.04	195.58	198.12	200.66
8	203.20	205.74	208.28	210.82	213.36	215.90	218.44	220.98	223.52	226.06
9	228.60	231.14	233.68	236.22	238.76	241.30	243.84	246.38	248.92	251.46
10	254.00	256.54	259.08	261.62	264.16	266.70	269.24	271.78	274.32	276.86
11	279.40	281.94	284.48	287.02	289.56	292.10	294.64	297.18	299.72	302.26
12	304.80	307.34	309.88	312.42	314.96	317.50	320.04	322.58	325.12	327.66
13	330.20	332.74	335.28	337.82	340.36	342.90	345.44	347.98	350.52	353.06
14	355.60	358.14	360.68	363.22	365.76	368.30	370.84	373.38	375.92	378.46
15	381.00	383.54	386.08	388.62	391.16	393.70	396.24	398.78	401.32	403.86
16	406.40	408.94	411.48	414.02	416.56	419.10	421.64	424.18	426.72	429.26
17	431.80	434.34	436.88	439.42	441.96	444.50	447.04	449.58	452.12	454.66
18	457.20	459.74	462.28	464.82	467.36	469.90	472.44	474.98	477.52	480.06
19	482.60	485.14	487.68	490.22	492.76	495.30	497.84	500.38	502.92	505.46
20	508.00	510.54	513.08	515.62	518.16	520.70	523.24	525.78	528.32	530.86
21	533.40	535.94	538.48	541.02	543.56	546.10	548.64	551.18	553.72	556.26
22	558.80	561.34	563.88	566.42	568.96	571.50	574.04	576.58	579.12	581.66
23	584.20	586.74	589.28	591.82	594.36	596.90	599.44	601.98	604.52	607.06
24	609.60	612.14	614.68	617.22	619.76	622.30	624.84	627.38	629.92	632.46
25	635.00	637.54	640.08	642.62	645.16	647.70	650.24	652.78	655.32	657.86
26	660.40	662.94	665.48	668.02	670.56	673.10	675.64	678.18	680.72	683.26
27	685.80	688.34	690.88	693.42	695.96	698.50	701.04	703.58	706.12	708.66
28	711.20	713.74	716.28	718.82	721.36	723.90	726.44	728.98	731.52	734.06
29	736.60	739.14	741.68	744.22	746.76	749.30	751.84	754.38	756.92	759.46
30	762.00	764.54	767.08	769.62	772.16	774.70	777.24	779.78	782.32	784.86
31	787.40	789.94	792.48	795.02	797.56	800.10	802.64	805.18	807.72	810.26
32	812.80	815.34	817.88	820.42	822.96	825.50	828.04	830.58	833.12	835.66
33	838.20	840.74	843.28	845.82	848.36	850.90	853.44	855.98	858.52	861.06
34	863.60	866.14	868.68	871.22	873.76	876.30	878.84	881.38	883.92	886.46
35	889.00	891.54	894.08	896.62	899.16	901.70	904.24	906.78	909.32	911.86
36	914.40	916.94	919.48	922.02	924.56	927.10	929.64	932.18	934.72	937.26
37	939.80	942.34	944.88	947.42	949.96	952.50	955.04	957.58	960.12	962.66
38	965.20	967.74	970.28	972.82	975.36	977.90	980.44	982.98	985.52	988.06
39	990.60	993.14	995.68	998.22	1000.76	1003.30	1005.84	1008.38	1010.92	1013.46

inches	0.0	0.1	0.2	0.3	0.4	0.5	0.6	0.7	0.8	0.9
40	1016.00	1018.54	1021.08	1023.62	1026.16	1028.70	1031.24	1033.78	1036.32	1038.86
41	1041.40	1043.94	1046.48	1049.02	1051.56	1054.10	1056.64	1059.18	1061.72	1064.26
42	1066.80	1069.34	1071.88	1074.42	1076.96	1079.50	1082.04	1084.58	1087.12	1089.66
43	1092.20	1094.74	1097.28	1099.82	1102.36	1104.90	1107.44	1109.98	1112.52	1115.06
44	1117.60	1120.14	1122.68	1125.22	1127.76	1130.30	1132.84	1135.38	1137.92	1140.46
45	1143.00	1145.54	1148.08	1150.62	1153.16	1155.70	1158.24	1160.78	1163.32	1165.86
46	1168.40	1170.94	1173.48	1176.02	1178.56	1181.10	1183.64	1186.18	1188.72	1191.26
47	1193.80	1196.34	1198.88	1201.42	1203.96	1206.50	1209.04	1211.58	1214.12	1216.66
48	1219.20	1221.74	1224.28	1226.82	1229.36	1231.90	1234.44	1236.98	1239.52	1242.06
49	1244.60	1247.14	1249.68	1252.22	1254.76	1257.30	1259.84	1262.38	1264.92	1267.46
50	1270.00	1272.54	1275.08	1277.62	1280.16	1282.70	1285.24	1287.78	1290.32	1292.86
51	1295.40	1297.94	1300.48	1303.02	1305.56	1308.10	1310.64	1313.18	1315.72	1318.26
52	1320.80	1323.34	1325.88	1328.42	1330.96	1333.50	1336.04	1338.58	1341.12	1343.66
53	1346.20	1348.74	1351.28	1353.82	1356.36	1358.90	1361.44	1363.98	1366.52	1369.06
54	1371.60	1374.14	1376.68	1379.22	1381.76	1384.30	1386.84	1389.38	1391.92	1394.46
55	1397.00	1399.54	1402.08	1404.62	1407.16	1409.70	1412.24	1414.78	1417.32	1419.86
56	1422.40	1424.94	1427.48	1430.02	1432.56	1435.10	1437.64	1440.18	1442.72	1445.26
57	1447.80	1450.34	1452.88	1455.42	1457.96	1460.50	1463.04	1465.58	1468.12	1470.66
58	1473.20	1475.74	1478.28	1480.82	1483.36	1485.90	1488.44	1490.98	1493.52	1496.06
59	1498.60	1501.14	1503.68	1506.22	1508.76	1511.30	1513.84	1516.38	1518.92	1521.46
60	1524.00	1526.54	1529.08	1531.62	1534.16	1536.70	1539.24	1541.78	1544.32	1546.86
61	1549.40	1551.94	1554.48	1557.02	1559.56	1562.10	1564.64	1567.18	1569.72	1572.26
62	1574.80	1577.34	1579.88	1582.42	1584.96	1587.50	1590.04	1592.58	1595.12	1597.66
63	1600.20	1602.74	1605.28	1607.82	1610.36	1612.90	1615.44	1617.98	1620.52	1623.06
64	1625.60	1628.14	1630.68	1633.22	1635.76	1638.30	1640.84	1643.38	1645.92	1648.46
65	1651.00	1653.54	1656.08	1658.62	1661.16	1663.70	1666.24	1668.78	1671.32	1673.86
66	1676.40	1678.94	1681.48	1684.02	1686.56	1689.10	1691.64	1694.18	1696.72	1699.26
67	1701.80	1704.34	1706.88	1709.42	1711.96	1714.50	1717.04	1719.58	1722.12	1724.66
68	1727.20	1729.74	1732.28	1734.82	1737.36	1739.90	1742.44	1744.98	1747.52	1750.06
69	1752.60	1755.14	1757.68	1760.22	1762.76	1765.30	1767.84	1770.38	1772.92	1775.46
70	1778.00	1780.54	1783.08	1785.62	1788.16	1790.70	1793.24	1795.78	1798.32	1800.86
71	1803.40	1805.94	1808.48	1811.02	1813.56	1816.10	1818.64	1821.18	1823.72	1826.26
72	1828.80	1831.34	1833.88	1836.42	1838.96	1841.50	1844.04	1846.58	1849.12	1851.66
73	1854.20	1856.74	1859.28	1861.82	1864.36	1866.90	1869.44	1871.98	1874.52	1877.06
74	1879.60	1882.14	1884.68	1887.22	1889.76	1892.30	1894.84	1897.38	1899.92	1902.46
75	1905.00	1907.54	1910.08	1912.62	1915.16	1917.70	1920.24	1922.78	1925.32	1927.86
76	1930.40	1932.94	1935.48	1938.02	1940.56	1943.10	1945.64	1948.18	1950.72	1953.26
77	1955.80	1958.34	1960.88	1963.42	1965.96	1968.50	1971.04	1973.58	1976.12	1978.66
78	1981.20	1983.74	1986.28	1988.82	1991.36	1993.90	1996.44	1998.98	2001.52	2004.06
79	2006.60	2009.14	2011.68	2014.22	2016.76	2019.30	2021.84	2024.38	2026.92	2029.46
80	2032.00	2034.54	2037.08	2039.62	2042.16	2044.70	2047.24	2049.78	2052.32	2054.86
81	2057.40	2059.94	2062.48	2065.02	2067.56	2070.10	2072.64	2075.18	2077.72	2080.26
82	2082.80	2085.34	2087.88	2090.42	2092.96	2095.50	2098.04	2100.58	2103.12	2105.66
83	2108.20	2110.74	2113.28	2115.82	2118.36	2120.90	2123.44	2125.98	2128.52	2131.06
84	2133.60	2136.14	2138.68	2141.22	2143.76	2146.30	2148.84	2151.38	2153.92	2156.46

inches	0.0	0.1	0.2	0.3	0.4	0.5	0.6	0.7	0.8	0.9
85	2159.00	2161.54	2164.08	2166.62	2169.16	2171.70	2174.24	2176.78	2179.32	2181.86
86	2184.40	2186.94	2189.48	2192.02	2194.56	2197.10	2199.64	2202.18	2204.72	2207.26
87	2209.80	2212.34	2214.88	2217.42	2219.96	2222.50	2225.04	2227.58	2230.12	2232.66
88	2235.20	2237.74	2240.28	2242.82	2245.36	2247.90	2250.44	2252.98	2255.52	2258.06
89	2260.60	2263.14	2265.68	2268.22	2270.76	2273.30	2275.84	2278.38	2280.92	2283.46
90	2286.00	2288.54	2291.08	2293.62	2296.16	2298.70	2301.24	2303.78	2306.32	2308.86
91	2311.40	2313.94	2316.48	2319.02	2321.56	2324.10	2326.64	2329.18	2331.72	2334.26
92	2336.80	2339.34	2341.88	2344.42	2346.96	2349.50	2352.04	2354.58	2357.12	2359.66
93	2362.20	2364.74	2367.28	2369.82	2372.36	2374.90	2377.44	2379.98	2382.52	2385.06
94	2387.60	2390.14	2392.68	2395.22	2397.76	2400.30	2402.84	2405.38	2407.92	2410.46
95	2413.00	2415.54	2418.08	2420.62	2423.16	2425.70	2428.24	2430.78	2433.32	2435.86
96	2438.40	2440.94	2443.48	2446.02	2448.56	2451.10	2453.64	2456.18	2458.72	2461.26
97	2463.80	2466.34	2468.88	2471.42	2473.96	2476.50	2479.04	2481.58	2484.12	2486.66
98	2489.20	2491.74	2494.28	2496.82	2499.36	2501.90	2504.44	2506.98	2509.52	2512.06
99	2514.60	2517.14	2519.68	2522.22	2524.76	2527.30	2529.84	2532.38	2534.92	2537.46

Note: In order to convert:
  (i) inches to centimetres, shift the decimal point in the above table one place to the left.
  (ii) inches to metres, shift the decimal point in the above table three places to the left.

Auxiliary Table—Fractions and Decimals of an Inch to Millimetres (exact values)

	inch	millimetre		inch	millimetre
$1/64$	0.015625	0.396875	$\frac{1}{8}$	0.125	3.175
$1/32$	0.03125	0.79375	$\frac{1}{4}$	0.25	6.35
$1/16$	0.0625	1.5875	$\frac{1}{2}$	0.5	12.7

Examples (i) | 5 ft $4\frac{1}{8}$ in = $64\frac{1}{8}$ in = 1625.6 + 3.175 mm = 1628.775 mm
         (ii) | 227.3 in = 5080.0 + 693.42 mm = 5773.42 mm

Table 2 MILLIMETRES TO INCHES

[mm to in]   1 mm = 0.039 370 08 in

(a)

millimetres	0	1000	2000	3000	4000	5000	6000	7000	8000	9000
0	—	39.37	78.74	118.11	157.48	196.85	236.22	275.59	314.96	354.33
10000	393.70	433.07	472.44	511.81	551.18	590.55	629.92	669.29	708.66	748.03

(b)

millimetres	0	1	2	3	4	5	6	7	8	9
0	—	0.04	0.08	0.12	0.16	0.20	0.24	0.28	0.31	0.35
10	0.39	0.43	0.47	0.51	0.55	0.59	0.63	0.67	0.71	0.75
20	0.79	0.83	0.87	0.91	0.94	0.98	1.02	1.06	1.10	1.14
30	1.18	1.22	1.26	1.30	1.34	1.38	1.42	1.46	1.50	1.54
40	1.57	1.61	1.65	1.69	1.73	1.77	1.81	1.85	1.89	1.93
50	1.97	2.01	2.05	2.09	2.13	2.17	2.20	2.24	2.28	2.32
60	2.36	2.40	2.44	2.48	2.52	2.56	2.60	2.64	2.68	2.72
70	2.76	2.80	2.83	2.87	2.91	2.95	2.99	3.03	3.07	3.11
80	3.15	3.19	3.23	3.27	3.31	3.35	3.39	3.43	3.46	3.50
90	3.54	3.58	3.62	3.66	3.70	3.74	3.78	3.82	3.86	3.90

millimetres	0	1	2	3	4	5	6	7	8	9
100	3.94	3.98	4.02	4.06	4.09	4.13	4.17	4.21	4.25	4.29
110	4.33	4.37	4.41	4.45	4.49	4.53	4.57	4.61	4.65	4.69
120	4.72	4.76	4.80	4.84	4.88	4.92	4.96	5.00	5.04	5.08
130	5.12	5.16	5.20	5.24	5.28	5.31	5.35	5.39	5.43	5.47
140	5.51	5.55	5.59	5.63	5.67	5.71	5.75	5.79	5.83	5.87
150	5.91	5.94	5.98	6.02	6.06	6.10	6.14	6.18	6.22	6.26
160	6.30	6.34	6.38	6.42	6.46	6.50	6.54	6.57	6.61	6.65
170	6.69	6.73	6.77	6.81	6.85	6.89	6.93	6.97	7.01	7.05
180	7.09	7.13	7.17	7.20	7.24	7.28	7.32	7.36	7.40	7.44
190	7.48	7.52	7.56	7.60	7.64	7.68	7.72	7.76	7.80	7.83
200	7.87	7.91	7.95	7.99	8.03	8.07	8.11	8.15	8.19	8.23
210	8.27	8.31	8.35	8.39	8.43	8.46	8.50	8.54	8.58	8.62
220	8.66	8.70	8.74	8.78	8.82	8.86	8.90	8.94	8.98	9.02
230	9.06	9.09	9.13	9.17	9.21	9.25	9.29	9.33	9.37	9.41
240	9.45	9.49	9.53	9.57	9.61	9.65	9.69	9.72	9.76	9.80
250	9.84	9.88	9.92	9.96	10.00	10.04	10.08	10.12	10.16	10.20
260	10.24	10.28	10.31	10.35	10.39	10.43	10.47	10.51	10.55	10.59
270	10.63	10.67	10.71	10.75	10.79	10.83	10.87	10.91	10.94	10.98
280	11.02	11.06	11.10	11.14	11.18	11.22	11.26	11.30	11.34	11.38
290	11.42	11.46	11.50	11.54	11.57	11.61	11.65	11.69	11.73	11.77
300	11.81	11.85	11.89	11.93	11.97	12.01	12.05	12.09	12.13	12.17
310	12.20	12.24	12.28	12.32	12.36	12.40	12.44	12.48	12.52	12.56
320	12.60	12.64	12.68	12.72	12.76	12.80	12.83	12.87	12.91	12.95
330	12.99	13.03	13.07	13.11	13.15	13.19	13.23	13.27	13.31	13.35
340	13.39	13.43	13.46	13.50	13.54	13.58	13.62	13.66	13.70	13.74
350	13.78	13.82	13.86	13.90	13.94	13.98	14.02	14.06	14.09	14.13
360	14.17	14.21	14.25	14.29	14.33	14.37	14.41	14.45	14.49	14.53
370	14.57	14.61	14.65	14.69	14.72	14.76	14.80	14.84	14.88	14.92
380	14.96	15.00	15.04	15.08	15.12	15.16	15.20	15.24	15.28	15.31
390	15.35	15.39	15.43	15.47	15.51	15.55	15.59	15.63	15.67	15.71
400	15.75	15.79	15.83	15.87	15.91	15.94	15.98	16.02	16.06	16.10
410	16.14	16.18	16.22	16.26	16.30	16.34	16.38	16.42	16.46	16.50
420	16.54	16.57	16.61	16.65	16.69	16.73	16.77	16.81	16.85	16.89
430	16.93	16.97	17.01	17.05	17.09	17.13	17.17	17.20	17.24	17.28
440	17.32	17.36	17.40	17.44	17.48	17.52	17.56	17.60	17.64	17.68
450	17.72	17.76	17.80	17.83	17.87	17.91	17.95	17.99	18.03	18.07
460	18.11	18.15	18.19	18.23	18.27	18.31	18.35	18.39	18.43	18.46
470	18.50	18.54	18.58	18.62	18.66	18.70	18.74	18.78	18.82	18.86
480	18.90	18.94	18.98	19.02	19.06	19.09	19.13	19.17	19.21	19.25
490	19.29	19.33	19.37	19.41	19.45	19.49	19.53	19.57	19.61	19.65
500	19.69	19.72	19.76	19.80	19.84	19.88	19.92	19.96	20.00	20.04
510	20.08	20.12	20.16	20.20	20.24	20.28	20.31	20.35	20.39	20.43
520	20.47	20.51	20.55	20.59	20.63	20.67	20.71	20.75	20.79	20.83
530	20.87	20.91	20.94	20.98	21.02	21.06	21.10	21.14	21.18	21.22
540	21.26	21.30	21.34	21.38	21.42	21.46	21.50	21.54	21.57	21.61

millimetres	0	1	2	3	4	5	6	7	8	9
550	21.65	21.69	21.73	21.77	21.81	21.85	21.89	21.93	21.97	22.01
560	22.05	22.09	22.13	22.17	22.20	22.24	22.28	22.32	22.36	22.40
570	22.44	22.48	22.52	22.56	22.60	22.64	22.68	22.72	22.76	22.80
580	22.83	22.87	22.91	22.95	22.99	23.03	23.07	23.11	23.15	23.19
590	23.23	23.27	23.31	23.35	23.39	23.43	23.46	23.50	23.54	23.58
600	23.62	23.66	23.70	23.74	23.78	23.82	23.86	23.90	23.94	23.98
610	24.02	24.06	24.09	24.13	24.17	24.21	24.25	24.29	24.33	24.37
620	24.41	24.45	24.49	24.53	24.57	24.61	24.65	24.69	24.72	24.76
630	24.80	24.84	24.88	24.92	24.96	25.00	25.04	25.08	25.12	25.16
640	25.20	25.24	25.28	25.31	25.35	25.39	25.43	25.47	25.51	25.55
650	25.59	25.63	25.67	25.71	25.75	25.79	25.83	25.87	25.91	25.94
660	25.98	26.02	26.06	26.10	26.14	26.18	26.22	26.26	26.30	26.34
670	26.38	26.42	26.46	26.50	26.54	26.57	26.61	26.65	26.69	26.73
680	26.77	26.81	26.85	26.89	26.93	26.97	27.01	27.05	27.09	27.13
690	27.17	27.20	27.24	27.28	27.32	27.36	27.40	27.44	27.48	27.52
700	27.56	27.60	27.64	27.68	27.72	27.76	27.80	27.83	27.87	27.91
710	27.95	27.99	28.03	28.07	28.11	28.15	28.19	28.23	28.27	28.31
720	28.35	28.39	28.43	28.46	28.50	28.54	28.58	28.62	28.66	28.70
730	28.74	28.78	28.82	28.86	28.90	28.94	28.98	29.02	29.06	29.09
740	29.13	29.17	29.21	29.25	29.29	29.33	29.37	29.41	29.45	29.49
750	29.53	29.57	29.61	29.65	29.69	29.72	29.76	29.80	29.84	29.88
760	29.92	29.96	30.00	30.04	30.08	30.12	30.16	30.20	30.24	30.28
770	30.31	30.35	30.39	30.43	30.47	30.51	30.55	30.59	30.63	30.67
780	30.71	30.75	30.79	30.83	30.87	30.91	30.94	30.98	31.02	31.06
790	31.10	31.14	31.18	31.22	31.26	31.30	31.34	31.38	31.42	31.46
800	31.50	31.54	31.57	31.61	31.65	31.69	31.73	31.77	31.81	31.85
810	31.89	31.93	31.97	32.01	32.05	32.09	32.13	32.17	32.20	32.24
820	32.28	32.32	32.36	32.40	32.44	32.48	32.52	32.56	32.60	32.64
830	32.68	32.72	32.76	32.80	32.83	32.87	32.91	32.95	32.99	33.03
840	33.07	33.11	33.15	33.19	33.23	33.27	33.31	33.35	33.39	33.43
850	33.46	33.50	33.54	33.58	33.62	33.66	33.70	33.74	33.78	33.82
860	33.86	33.90	33.94	33.98	34.02	34.06	34.09	34.13	34.17	34.21
870	34.25	34.29	34.33	34.37	34.41	34.45	34.49	34.53	34.57	34.61
880	34.65	34.69	34.72	34.76	34.80	34.84	34.88	34.92	34.96	35.00
890	35.04	35.08	35.12	35.16	35.20	35.24	35.28	35.31	35.35	35.39
900	35.43	35.47	35.51	35.55	35.59	35.63	35.67	35.71	35.75	35.79
910	35.83	35.87	35.91	35.94	35.98	36.02	36.06	36.10	36.14	36.18
920	36.22	36.26	36.30	36.34	36.38	36.42	36.46	36.50	36.54	36.57
930	36.61	36.65	36.69	36.73	36.77	36.81	36.85	36.89	36.93	36.97
940	37.01	37.05	37.09	37.13	37.17	37.20	37.24	37.28	37.32	37.36
950	37.40	37.44	37.48	37.52	37.56	37.60	37.64	37.68	37.72	37.76
960	37.80	37.83	37.87	37.91	37.95	37.99	38.03	38.07	38.11	38.15
870	38.19	38.23	38.27	38.31	38.35	38.39	38.43	38.46	38.50	38.54
980	38.58	38.62	38.66	38.70	38.74	38.78	38.82	38.86	38.90	38.94
990	38.98	39.02	39.06	39.09	39.13	39.17	39.21	39.25	39.29	39.33

Example: 13872 mm = 511.81 + 34.33 in = 546.14 in

13

Table 3 FEET TO METRES

[ft to m]   1 ft = 0.304 8 m (exactly)
All values in Tables 3(a) and (b) are exact.

(a)

feet	0	1000	2000	3000	4000	5000	6000	7000	8000	9000
0	—	304.8	609.6	914.4	1219.2	1524.0	1828.8	2133.6	2438.4	2743.2
10000	3048.0	3352.8	3657.6	3962.4	4267.2	4572.0	4876.8	5181.6	5486.4	5791.2
20000	6096.0	6400.8	6705.6	7010.4	7315.2	7620.0	7924.8	8229.6	8534.4	8839.2

(b)

feet	0	1	2	3	4	5	6	7	8	9
0	—	0.3048	0.6096	0.9144	1.2192	1.5240	1.8288	2.1336	2.4384	2.7432
10	3.0480	3.3528	3.6576	3.9624	4.2672	4.5720	4.8768	5.1816	5.4864	5.7912
20	6.0960	6.4008	6.7056	7.0104	7.3152	7.6200	7.9248	8.2296	8.5344	8.8392
30	9.1440	9.4488	9.7536	10.0584	10.3632	10.6680	10.9728	11.2776	11.5824	11.8872
40	12.1920	12.4968	12.8016	13.1064	13.4112	13.7160	14.0208	14.3256	14.6304	14.9352
50	15.2400	15.5448	15.8496	16.1544	16.4592	16.7640	17.0688	17.3736	17.6784	17.9832
60	18.2880	18.5928	18.8976	19.2024	19.5072	19.8120	20.1168	20.4216	20.7264	21.0312
70	21.3360	21.6408	21.9456	22.2504	22.5552	22.8600	23.1648	23.4696	23.7744	24.0792
80	24.3840	24.6888	24.9936	25.2984	25.6032	25.9080	26.2128	26.5176	26.8224	27.1272
90	27.4320	27.7368	28.0416	28.3464	28.6512	28.9560	29.2608	29.5656	29.8704	30.1752
100	30.4800	30.7848	31.0896	31.3944	31.6992	32.0040	32.3088	32.6136	32.9184	33.2232
110	33.5280	33.8328	34.1376	34.4424	34.7472	35.0520	35.3568	35.6616	35.9664	36.2712
120	36.5760	36.8808	37.1856	37.4904	37.7952	38.1000	38.4048	38.7096	39.0144	39.3192
130	39.6240	39.9288	40.2336	40.5384	40.8432	41.1480	41.4528	41.7576	42.0624	42.3672
140	42.6720	42.9768	43.2816	43.5864	43.8912	44.1960	44.5008	44.8056	45.1104	45.4152
150	45.7200	46.0248	46.3296	46.6344	46.9392	47.2440	47.5488	47.8536	48.1584	48.4632
160	48.7680	49.0728	49.3776	49.6824	49.9872	50.2920	50.5968	50.9016	51.2064	51.5112
170	51.8160	52.1208	52.4256	52.7304	53.0352	53.3400	53.6448	53.9496	54.2544	54.5592
180	54.8640	55.1688	55.4736	55.7784	56.0832	56.3880	56.6928	56.9976	57.3024	57.6072
190	57.9120	58.2168	58.5216	58.8264	59.1312	59.4360	59.7408	60.0456	60.3504	60.6552
200	60.9600	61.2648	61.5696	61.8744	62.1792	62.4840	62.7888	63.0936	63.3984	63.7032
210	64.0080	64.3128	64.6176	64.9224	65.2272	65.5320	65.8368	66.1416	66.4464	66.7512
220	67.0560	67.3608	67.6656	67.9704	68.2752	68.5800	68.8848	69.1896	69.4944	69.7992
230	70.1040	70.4088	70.7136	71.0184	71.3232	71.6280	71.9328	72.2376	72.5424	72.8472
240	73.1520	73.4568	73.7616	74.0664	74.3712	74.6760	74.9808	75.2856	75.5904	75.8952
250	76.2000	76.5048	76.8096	77.1144	77.4192	77.7240	78.0288	78.3336	78.6384	78.9432
260	79.2480	79.5528	79.8576	80.1624	80.4672	80.7720	81.0768	81.3816	81.6864	81.9912
270	82.2960	82.6008	82.9056	83.2104	83.5152	83.8200	84.1248	84.4296	84.7344	85.0392
280	85.3440	85.6488	85.9536	86.2584	86.5632	86.8680	87.1728	87.4776	87.7824	88.0872
290	88.3920	88.6968	89.0016	89.3064	89.6112	89.9160	90.2208	90.5256	90.8304	91.1352
300	91.4400	91.7448	92.0496	92.3544	92.6592	92.9640	93.2688	93.5736	93.8784	94.1832
310	94.4880	94.7928	95.0976	95.4024	95.7072	96.0120	96.3168	96.6216	96.9264	97.2312
320	97.5360	97.8408	98.1456	98.4504	98.7552	99.0600	99.3648	99.6696	99.9744	100.2792
330	100.5840	100.8888	101.1936	101.4984	101.8032	102.1080	102.4128	102.7176	103.0224	103.3272
340	103.6320	103.9368	104.2416	104.5464	104.8512	105.1560	105.4608	105.7656	106.0704	106.3752

feet	0	1	2	3	4	5	6	7	8	9
350	106.6800	106.9848	107.2896	107.5944	107.8992	108.2040	108.5088	108.8136	109.1184	109.4232
360	109.7280	110.0328	110.3376	110.6424	110.9472	111.2520	111.5568	111.8616	112.1664	112.4712
370	112.7760	113.0808	113.3856	113.6904	113.9952	114.3000	114.6048	114.9096	115.2144	115.5192
380	115.8240	116.1288	116.4336	116.7384	117.0432	117.3480	117.6528	117.9576	118.2624	118.5672
390	118.8720	119.1768	119.4816	119.7864	120.0912	120.3960	120.7008	121.0056	121.3104	121.6152
400	121.9200	122.2248	122.5296	122.8344	123.1392	123.4440	123.7488	124.0536	124.3584	124.6632
410	124.9680	125.2728	125.5776	125.8824	126.1872	126.4920	126.7968	127.1016	127.4064	127.7112
420	128.0160	128.3208	128.6256	128.9304	129.2352	129.5400	129.8448	130.1496	130.4544	130.7592
430	131.0640	131.3688	131.6736	131.9784	132.2832	132.5880	132.8928	133.1976	133.5024	133.8072
440	134.1120	134.4168	134.7216	135.0264	135.3312	135.6360	135.9408	136.2456	136.5504	136.8552
450	137.1600	137.4648	137.7696	138.0744	138.3792	138.6840	138.9888	139.2936	139.5984	139.9032
460	140.2080	140.5128	140.8176	141.1224	141.4272	141.7320	142.0368	142.3416	142.6464	142.9512
470	143.2560	143.5608	143.8656	144.1704	144.4752	144.7800	145.0848	145.3896	145.6944	145.9992
480	146.3040	146.6088	146.9136	147.2184	147.5232	147.8280	148.1328	148.4376	148.7424	149.0472
490	149.3520	149.6568	149.9616	150.2664	150.5712	150.8760	151.1808	151.4856	151.7904	152.0952
500	152.4000	152.7048	153.0096	153.3144	153.6192	153.9240	154.2288	154.5336	154.8384	155.1432
510	155.4480	155.7528	156.0576	156.3624	156.6672	156.9720	157.2768	157.5816	157.8864	158.1912
520	158.4960	158.8008	159.1056	159.4104	159.7152	160.0200	160.3248	160.6296	160.9344	161.2392
530	161.5440	161.8488	162.1536	162.4584	162.7632	163.0680	163.3728	163.6776	163.9824	164.2872
540	164.5920	164.8968	165.2016	165.5064	165.8112	166.1160	166.4208	166.7256	167.0304	167.3352
550	167.6400	167.9448	168.2496	168.5544	168.8592	169.1640	169.4688	169.7736	170.0784	170.3832
560	170.6880	170.9928	171.2976	171.6024	171.9072	172.2120	172.5168	172.8216	173.1264	173.4312
570	173.7360	174.0408	174.3456	174.6504	174.9552	175.2600	175.5648	175.8696	176.1744	176.4792
580	176.7840	177.0888	177.3936	177.6984	178.0032	178.3080	178.6128	178.9176	179.2224	179.5272
590	179.8320	180.1368	180.4416	180.7464	181.0512	181.3560	181.6608	181.9656	182.2704	182.5752
600	182.8800	183.1848	183.4896	183.7944	184.0992	184.4040	184.7088	185.0136	185.3184	185.6232
610	185.9280	186.2328	186.5376	186.8424	187.1472	187.4520	187.7568	188.0616	188.3664	188.6712
620	188.9760	189.2808	189.5856	189.8904	190.1952	190.5000	190.8048	191.1096	191.4144	191.7192
630	192.0240	192.3288	192.6336	192.9384	193.2432	193.5480	193.8528	194.1576	194.4624	194.7672
640	195.0720	195.3768	195.6816	195.9864	196.2912	196.5960	196.9008	197.2056	197.5104	197.8152
650	198.1200	198.4248	198.7296	199.0344	199.3392	199.6440	199.9488	200.2536	200.5584	200.8632
660	201.1680	201.4728	201.7776	202.0824	202.3872	202.6920	202.9968	203.3016	203.6064	203.9112
670	204.2160	204.5208	204.8256	205.1304	205.4352	205.7400	206.0448	206.3496	206.6544	206.9592
680	207.2640	207.5688	207.8736	208.1784	208.4832	208.7880	209.0928	209.3976	209.7024	210.0072
690	210.3120	210.6168	210.9216	211.2264	211.5312	211.8360	212.1408	212.4456	212.7504	213.0552
700	213.3600	213.6648	213.9696	214.2744	214.5792	214.8840	215.1888	215.4936	215.7984	216.1032
710	216.4080	216.7128	217.0176	217.3224	217.6272	217.9320	218.2368	218.5416	218.8464	219.1512
720	219.4560	219.7608	220.0656	220.3704	220.6752	220.9800	221.2848	221.5896	221.8944	222.1992
730	222.5040	222.8088	223.1136	223.4184	223.7232	224.0280	224.3328	224.6376	224.9424	225.2472
740	225.5520	225.8568	226.1616	226.4664	226.7712	227.0760	227.3808	227.6856	227.9904	228.2952
750	228.6000	228.9048	229.2096	229.5144	229.8192	230.1240	230.4288	230.7336	231.0384	231.3432
760	231.6480	231.9528	232.2576	232.5624	232.8672	233.1720	233.4768	233.7816	234.0864	234.3912
770	234.6960	235.0008	235.3056	235.6104	235.9152	236.2200	236.5248	236.8296	237.1344	237.4392
780	237.7440	238.0488	238.3536	238.6584	238.9632	239.2680	239.5728	239.8776	240.1824	240.4872
790	240.7920	241.0968	241.4016	241.7064	242.0112	242.3160	242.6208	242.9256	243.2304	243.5352

feet	0	1	2	3	4	5	6	7	8	9
800	243.8400	244.1448	244.4496	244.7544	245.0592	245.3640	245.6688	245.9736	246.2784	246.5832
810	246.8880	247.1928	247.4976	247.8024	248.1072	248.4120	248.7168	249.0216	249.3264	249.6312
820	249.9360	250.2408	250.5456	250.8504	251.1552	251.4600	251.7648	252.0696	252.3744	252.6792
830	252.9840	253.2888	253.5936	253.8984	254.2032	254.5080	254.8128	255.1176	255.4224	255.7272
840	256.0320	256.3368	256.6416	256.9464	257.2512	257.5560	257.8608	258.1656	258.4704	258.7752
850	259.0800	259.3848	259.6896	259.9944	260.2992	260.6040	260.9088	261.2136	261.5184	261.8232
860	262.1280	262.4328	262.7376	263.0424	263.3472	263.6520	263.9568	264.2616	264.5664	264.8712
870	265.1760	265.4808	265.7856	266.0904	266.3952	266.7000	267.0048	267.3096	267.6144	267.9192
880	268.2240	268.5288	268.8336	269.1384	269.4432	269.7480	270.0528	270.3576	270.6624	270.9672
890	271.2720	271.5768	271.8816	272.1864	272.4912	272.7960	273.1008	273.4056	273.7104	274.0152
900	274.3200	274.6248	274.9296	275.2344	275.5392	275.8440	276.1488	276.4536	276.7584	277.0632
910	277.3680	277.6728	277.9776	278.2824	278.5872	278.8920	279.1968	279.5016	279.8064	280.1112
920	280.4160	280.7208	281.0256	281.3304	281.6352	281.9400	282.2448	282.5496	282.8544	283.1592
930	283.4640	283.7688	284.0736	284.3784	284.6832	284.9880	285.2928	285.5976	285.9024	286.2072
940	286.5120	286.8168	287.1216	287.4264	287.7312	288.0360	288.3408	288.6456	288.9504	289.2552
950	289.5600	289.8648	290.1696	290.4744	290.7792	291.0840	291.3888	291.6936	291.9984	292.3032
960	292.6080	292.9128	293.2176	293.5224	293.8272	294.1320	294.4368	294.7416	295.0464	295.3512
970	295.6560	295.9608	296.2656	296.5704	296.8752	297.1800	297.4848	297.7896	298.0944	298.3992
980	298.7040	299.0088	299.3136	299.6184	299.9232	300.2280	300.5328	300.8376	301.1424	301.4472
990	301.7520	302.0568	302.3616	302.6664	302.9712	303.2760	303.5808	303.8856	304.1904	304.4952

Note: The above table can be used to calculate:
(i)   velocity and (ii) acceleration:
    for example: 851 ft = 259.3848 m
    hence   851 ft s^{-1} = 259.3848 m s^{-1}
    and   851 ft s^{-2} = 259.3848 m s^{-2}
Example: 5486 ft s^{-1} = 1524.00 + 148.1328 m s^{-1} = 1672.1328 m s^{-1}

Table 4   METRES TO FEET
[m to ft]   1 m = 3.280 839 895 ft

(a)

metres	0	1000	2000	3000	4000	5000	6000	7000	8000	9000
0	—	3280.840	6561.680	9842.520	13123.360	16404.199	19685.039	22965.879	26246.719	29527.559
10000	32808.399	36089.239	39370.079	42650.919	45931.759	49212.598	52493.438	55774.278	59055.118	62335.958
20000	65616.798	68897.638	72178.478	75459.318	78740.157	82020.997	85301.837	88582.677	91863.517	95144.357

(b)

metres	0	1	2	3	4	5	6	7	8	9
0	—	3.281	6.562	9.843	13.123	16.404	19.685	22.966	26.247	29.528
10	32.808	36.089	39.370	42.651	45.932	49.213	52.493	55.774	59.055	62.336
20	65.617	68.898	72.178	75.459	78.740	82.021	85.302	88.583	91.864	95.144
30	98.425	101.706	104.987	108.268	111.549	114.829	118.110	121.391	124.672	127.953
40	131.234	134.514	137.795	141.076	144.357	147.638	150.919	154.199	157.480	160.761
50	164.042	167.323	170.604	173.885	177.165	180.446	183.727	187.008	190.289	193.570
60	196.850	200.131	203.412	206.693	209.974	213.255	216.535	219.816	223.097	226.378
70	229.659	232.940	236.220	239.501	242.782	246.063	249.344	252.625	255.906	259.186
80	262.467	265.748	269.029	272.310	275.591	278.871	282.152	285.433	288.714	291.995
90	295.276	298.556	301.837	305.118	308.399	311.680	314.961	318.241	321.522	324.803

metres	0	1	2	3	4	5	6	7	8	9
100	328.084	331.365	334.646	337.927	341.207	344.488	347.769	351.050	354.331	357.612
110	360.892	364.173	367.454	370.735	374.016	377.297	380.577	383.858	387.139	390.420
120	393.701	396.982	400.262	403.543	406.824	410.105	413.386	416.667	419.948	423.228
130	426.509	429.790	433.071	436.352	439.633	442.913	446.194	449.475	452.756	456.037
140	459.318	462.598	465.879	469.160	472.441	475.722	479.003	482.283	485.564	488.845
150	492.126	495.407	498.688	501.969	505.249	508.530	511.811	515.092	518.373	521.654
160	524.934	528.215	531.496	534.777	538.058	541.339	544.619	547.900	551.181	554.462
170	557.743	561.024	564.304	567.585	570.866	574.147	577.428	580.709	583.990	587.270
180	590.551	593.832	597.113	600.394	603.675	606.955	610.236	613.517	616.798	620.079
190	623.360	626.640	629.921	633.202	636.483	639.764	643.045	646.325	649.606	652.887
200	656.168	659.449	662.730	666.010	669.291	672.572	675.853	679.134	682.415	685.696
210	688.976	692.257	695.538	698.819	702.100	705.381	708.661	711.942	715.223	718.504
220	721.785	725.066	728.346	731.627	734.908	738.189	741.470	744.751	748.031	751.312
230	754.593	757.874	761.155	764.436	767.717	770.997	774.278	777.559	780.840	784.121
240	787.402	790.682	793.963	797.244	800.525	803.806	807.087	810.367	813.648	816.929
250	820.210	823.491	826.772	830.052	833.333	836.614	839.895	843.176	846.457	849.738
260	853.018	856.299	859.580	862.861	866.142	869.423	872.703	875.984	879.265	882.546
270	885.827	889.108	892.388	895.669	898.950	902.231	905.512	908.793	912.073	915.354
280	918.635	921.916	925.197	928.478	931.759	935.039	938.320	941.601	944.882	948.163
290	951.444	954.724	958.005	961.286	964.567	967.848	971.129	974.409	977.690	980.971
300	984.252	987.533	990.814	994.094	997.375	1000.656	1003.937	1007.218	1010.499	1013.780
310	1017.060	1020.341	1023.622	1026.903	1030.184	1033.465	1036.745	1040.026	1043.307	1046.588
320	1049.869	1053.150	1056.430	1059.711	1062.992	1066.273	1069.554	1072.835	1076.115	1079.396
330	1082.677	1085.958	1089.239	1092.520	1095.801	1099.081	1102.362	1105.643	1108.924	1112.205
340	1115.486	1118.766	1122.047	1125.328	1128.609	1131.890	1135.171	1138.451	1141.732	1145.013
350	1148.294	1151.575	1154.856	1158.136	1161.417	1164.698	1167.979	1171.260	1174.541	1177.822
360	1181.102	1184.383	1187.664	1190.945	1194.226	1197.507	1200.787	1204.068	1207.349	1210.630
370	1213.911	1217.192	1220.472	1223.753	1227.034	1230.315	1233.596	1236.877	1240.157	1243.438
380	1246.719	1250.000	1253.281	1256.562	1259.843	1263.123	1266.404	1269.685	1272.966	1276.247
390	1279.528	1282.808	1286.089	1289.370	1292.651	1295.932	1299.213	1302.493	1305.774	1309.055
400	1312.336	1315.617	1318.898	1322.178	1325.459	1328.740	1332.021	1335.302	1338.583	1341.864
410	1345.144	1348.425	1351.706	1354.987	1358.268	1361.549	1364.829	1368.110	1371.391	1374.672
420	1377.953	1381.234	1384.514	1387.795	1391.076	1394.357	1397.638	1400.919	1404.199	1407.480
430	1410.761	1414.042	1417.323	1420.604	1423.885	1427.165	1430.446	1433.727	1437.008	1440.289
440	1443.570	1446.850	1450.131	1453.412	1456.693	1459.974	1463.255	1466.535	1469.816	1473.097
450	1476.378	1479.659	1482.940	1486.220	1489.501	1492.782	1496.063	1499.344	1502.625	1505.906
460	1509.186	1512.467	1515.748	1519.029	1522.310	1525.591	1528.871	1532.152	1535.433	1538.714
470	1541.995	1545.276	1548.556	1551.837	1555.118	1558.399	1561.680	1564.961	1568.241	1571.522
480	1574.803	1578.084	1581.365	1584.646	1587.927	1591.207	1594.488	1597.769	1601.050	1604.331
490	1607.612	1610.892	1614.173	1617.454	1620.735	1624.016	1627.297	1630.577	1633.858	1637.139
500	1640.420	1643.701	1646.982	1650.262	1653.543	1656.824	1660.105	1663.386	1666.667	1669.948
510	1673.228	1676.509	1679.790	1683.071	1686.352	1689.633	1692.913	1696.194	1699.475	1702.756
520	1706.037	1709.318	1712.598	1715.879	1719.160	1722.441	1725.722	1729.003	1732.283	1735.564
530	1738.845	1742.126	1745.407	1748.688	1751.969	1755.249	1758.530	1761.811	1765.092	1768.373
540	1771.654	1774.934	1778.215	1781.496	1784.777	1788.058	1791.339	1794.619	1797.900	1801.181

metres	0	1	2	3	4	5	6	7	8	9
550	1804.462	1807.743	1811.024	1814.304	1817.585	1820.866	1824.147	1827.428	1830.709	1833.990
560	1837.270	1840.551	1843.832	1847.113	1850.394	1853.675	1856.955	1860.236	1863.517	1866.798
570	1870.079	1873.360	1876.640	1879.921	1883.202	1886.483	1889.764	1893.045	1896.325	1899.606
580	1902.887	1906.168	1909.449	1912.730	1916.010	1919.291	1922.572	1925.853	1929.134	1932.415
590	1935.696	1938.976	1942.257	1945.538	1948.819	1952.100	1955.381	1958.661	1961.942	1965.223
600	1968.504	1971.785	1975.066	1978.346	1981.627	1984.908	1988.189	1991.470	1994.751	1998.031
610	2001.312	2004.593	2007.874	2011.155	2014.436	2017.717	2020.997	2024.278	2027.559	2030.840
620	2034.121	2037.402	2040.682	2043.963	2047.244	2050.525	2053.806	2057.087	2060.367	2063.648
630	2066.929	2070.210	2073.491	2076.772	2080.052	2083.333	2086.614	2089.895	2093.176	2096.457
640	2099.738	2103.018	2106.299	2109.580	2112.861	2116.142	2119.423	2122.703	2125.984	2129.265
650	2132.546	2135.827	2139.108	2142.388	2145.669	2148.950	2152.231	2155.512	2158.793	2162.073
660	2165.354	2168.635	2171.916	2175.197	2178.478	2181.759	2185.039	2188.320	2191.601	2194.882
670	2198.163	2201.444	2204.724	2208.005	2211.286	2214.567	2217.848	2221.129	2224.409	2227.690
680	2230.971	2234.252	2237.533	2240.814	2244.094	2247.375	2250.656	2253.937	2257.218	2260.499
690	2263.780	2267.060	2270.341	2273.622	2276.903	2280.184	2283.465	2286.745	2290.026	2293.307
700	2296.588	2299.869	2303.150	2306.430	2309.711	2312.992	2316.273	2319.554	2322.835	2326.115
710	2329.396	2332.677	2335.958	2339.239	2342.520	2345.801	2349.081	2352.362	2355.643	2358.924
720	2362.205	2365.486	2368.766	2372.047	2375.328	2378.609	2381.890	2385.171	2388.451	2391.732
730	2395.013	2398.294	2401.575	2404.856	2408.136	2411.417	2414.698	2417.979	2421.260	2424.541
740	2427.822	2431.102	2434.383	2437.664	2440.945	2444.226	2447.507	2450.787	2454.068	2457.349
750	2460.630	2463.911	2467.192	2470.472	2473.753	2477.034	2480.315	2483.596	2486.877	2490.157
760	2493.438	2496.719	2500.000	2503.281	2506.562	2509.843	2513.123	2516.404	2519.685	2522.966
770	2526.247	2529.528	2532.808	2536.089	2539.370	2542.651	2545.932	2549.213	2552.493	2555.774
780	2559.055	2562.336	2565.617	2568.898	2572.178	2575.459	2578.740	2582.021	2585.302	2588.583
790	2591.864	2595.144	2598.425	2601.706	2604.987	2608.268	2611.549	2614.829	2618.110	2621.391
800	2624.672	2627.953	2631.234	2634.514	2637.795	2641.076	2644.357	2647.638	2650.919	2654.199
810	2657.480	2660.761	2664.042	2667.323	2670.604	2673.885	2677.165	2680.446	2683.727	2687.008
820	2690.289	2693.570	2696.850	2700.131	2703.412	2706.693	2709.974	2713.255	2716.535	2719.816
830	2723.097	2726.378	2729.659	2732.940	2736.220	2739.501	2742.782	2746.063	2749.344	2752.625
840	2755.906	2759.186	2762.467	2765.748	2769.029	2772.310	2775.591	2778.871	2782.152	2785.433
850	2788.714	2791.995	2795.276	2798.556	2801.837	2805.118	2808.399	2811.680	2814.961	2818.241
860	2821.522	2824.803	2828.084	2831.365	2834.646	2837.927	2841.207	2844.488	2847.769	2851.050
870	2854.331	2857.612	2860.892	2864.173	2867.454	2870.735	2874.016	2877.297	2880.577	2883.858
880	2887.139	2890.420	2893.701	2896.982	2900.262	2903.543	2906.824	2910.105	2913.386	2916.667
890	2919.948	2923.228	2926.509	2929.790	2933.071	2936.352	2939.633	2942.913	2946.194	2949.475
900	2952.756	2956.037	2959.318	2962.598	2965.879	2969.160	2972.441	2975.722	2979.003	2982.283
910	2985.564	2988.845	2992.126	2995.407	2998.688	3001.969	3005.249	3008.530	3011.811	3015.092
920	3018.373	3021.654	3024.934	3028.215	3031.496	3034.777	3038.058	3041.339	3044.619	3047.900
930	3051.181	3054.462	3057.743	3061.024	3064.304	3067.585	3070.866	3074.147	3077.428	3080.709
940	3083.990	3087.270	3090.551	3093.832	3097.113	3100.394	3103.675	3106.955	3110.236	3113.517
950	3116.798	3120.079	3123.360	3126.640	3129.921	3133.202	3136.483	3139.764	3143.045	3146.325
960	3149.606	3152.887	3156.168	3159.449	3162.730	3166.010	3169.291	3172.572	3175.853	3179.134
970	3182.415	3185.696	3188.976	3192.257	3195.538	3198.819	3202.100	3205.381	3208.661	3211.942
980	3215.223	3218.504	3221.785	3225.066	3228.346	3231.627	3234.908	3238.189	3241.470	3244.751
990	3248.031	3251.312	3254.593	3257.874	3261.155	3264.436	3267.717	3270.997	3274.278	3277.559

See Note at end of Table 3
Example: 8848 m = 26246.719 + 2782.152 ft = 29028.871 ft

Table 5 YARDS TO METRES

[yd to m]   1 yd = 0.914 4 m (exactly)
All values in Tables 5(a) and 5(b) are exact.

(a)

yards	0	100	200	300	400	500	600	700	800	900
0	—	91.44	182.88	274.32	365.76	457.20	548.64	640.08	731.52	822.96
1000	914.40	1005.84	1097.28	1188.72	1280.16	1371.60	1463.04	1554.48	1645.92	1737.36

(b)

yards	0	1	2	3	4	5	6	7	8	9
0	—	0.9144	1.8288	2.7432	3.6576	4.5720	5.4864	6.4008	7.3152	8.2296
10	9.1440	10.0584	10.9728	11.8872	12.8016	13.7160	14.6304	15.5448	16.4592	17.3736
20	18.2880	19.2024	20.1168	21.0312	21.9456	22.8600	23.7744	24.6888	25.6032	26.5176
30	27.4320	28.3464	29.2608	30.1752	31.0896	32.0040	32.9184	33.8328	34.7472	35.6616
40	36.5760	37.4904	38.4048	39.3192	40.2336	41.1480	42.0624	42.9768	43.8912	44.8056
50	45.7200	46.6344	47.5488	48.4632	49.3776	50.2920	51.2064	52.1208	53.0352	53.9496
60	54.8640	55.7784	56.6928	57.6072	58.5216	59.4360	60.3504	61.2648	62.1792	63.0936
70	64.0080	64.9224	65.8368	66.7512	67.6656	68.5800	69.4944	70.4088	71.3232	72.2376
80	73.1520	74.0664	74.9808	75.8952	76.8096	77.7240	78.6384	79.5528	80.4672	81.3816
90	82.2960	83.2104	84.1248	85.0392	85.9536	86.8680	87.7824	88.6968	89.6112	90.5256

Example: 1760 yd = 1554.48 + 54.864 m = 1609.344 m

Table 6 METRES TO YARDS

[m to yd]   1 m = 1.093 613 3 yd

(a)

metres	0	100	200	300	400	500	600	700	800	900
0	—	109.3613	218.7227	328.0840	437.4453	546.8067	656.1680	765.5293	874.8906	984.2520
1000	1093.6133	1202.9746	1312.3360	1421.6973	1531.0586	1640.4200	1749.7813	1859.1426	1968.5039	2077.8653

(b)

metres	0	1	2	3	4	5	6	7	8	9
0	—	1.0936	2.1872	3.2808	4.3745	5.4681	6.5617	7.6553	8.7489	9.8425
10	10.9361	12.0297	13.1234	14.2170	15.3106	16.4042	17.4978	18.5914	19.6850	20.7787
20	21.8723	22.9659	24.0595	25.1531	26.2467	27.3403	28.4339	29.5276	30.6212	31.7148
30	32.8084	33.9020	34.9956	36.0892	37.1829	38.2765	39.3701	40.4637	41.5573	42.6509
40	43.7445	44.8381	45.9318	47.0254	48.1190	49.2126	50.3062	51.3998	52.4934	53.5871
50	54.6807	55.7743	56.8679	57.9615	59.0551	60.1487	61.2423	62.3360	63.4296	64.5232
60	65.6168	66.7104	67.8040	68.8976	69.9913	71.0849	72.1785	73.2721	74.3657	75.4593
70	76.5529	77.6465	78.7402	79.8338	80.9274	82.0210	83.1146	84.2082	85.3018	86.3955
80	87.4891	88.5827	89.6763	90.7699	91.8635	92.9571	94.0507	95.1444	96.2380	97.3316
90	98.4252	99.5188	100.6124	101.7060	102.7997	103.8933	104.9869	106.0805	107.1741	108.2677

Example: 595 m = 546.8067 + 103.8933 yd = 650.7000 yd

Table 7 FATHOMS TO METRES

1 fathom = 1.828 8 m exactly
All values in Table 7(a) are exact

(a)

fathoms	0	100	200	300	400	500	600	700	800	900
0	—	182.88	365.76	548.64	731.52	914.40	1097.28	1280.16	1463.04	1645.92
1000	1828.80	2011.68	2194.56	2377.44	2560.32	2743.20	2926.08	3108.96	3291.84	3474.72

(b)

fathoms	0	1	2	3	4	5	6	7	8	9
0	—	1.83	3.66	5.49	7.32	9.14	10.97	12.80	14.63	16.46
10	18.29	20.12	21.95	23.77	25.60	27.43	29.26	31.09	32.92	34.75
20	36.58	38.40	40.23	42.06	43.89	45.72	47.55	49.38	51.21	53.04
30	54.86	56.69	58.52	60.35	62.18	64.01	65.84	67.67	69.49	71.32
40	73.15	74.98	76.81	78.64	80.47	82.30	84.12	85.95	87.78	89.61
50	91.44	93.27	95.10	96.93	98.76	100.58	102.41	104.24	106.07	107.90
60	109.73	111.56	113.39	115.21	117.04	118.87	120.70	122.53	124.36	126.19
70	128.02	129.84	131.67	133.50	135.33	137.16	138.99	140.82	142.65	144.48
80	146.30	148.13	149.96	151.79	153.62	155.45	157.28	159.11	160.93	162.76
90	164.59	166.42	168.25	170.08	171.91	173.74	175.56	177.39	179.22	181.05

Example: 1234 fathoms = 2194.56 + 62.18 m = 2256.74 m

Table 8 METRES TO FATHOMS

1 m = 0.546 807 fathom

(a)

metres	0	100	200	300	400	500	600	700	800	900
0	—	54.68	109.36	164.04	218.72	273.40	328.08	382.76	437.45	492.13
1000	546.81	601.49	656.17	710.85	765.53	820.21	874.89	929.57	984.25	1038.93

(b)

metres	0	1	2	3	4	5	6	7	8	9
0	—	0.55	1.09	1.64	2.19	2.73	3.28	3.83	4.37	4.92
10	5.47	6.01	6.56	7.11	7.66	8.20	8.75	9.30	9.84	10.39
20	10.94	11.48	12.03	12.58	13.12	13.67	14.22	14.76	15.31	15.86
30	16.40	16.95	17.50	18.04	18.59	19.14	19.69	20.23	20.78	21.33
40	21.87	22.42	22.97	23.51	24.06	24.61	25.15	25.70	26.25	26.79
50	27.34	27.89	28.43	28.98	29.53	30.07	30.62	31.17	31.71	32.26
60	32.81	33.36	33.90	34.45	35.00	35.54	36.09	36.64	37.18	37.73
70	38.28	38.82	39.37	39.92	40.46	41.01	41.56	42.10	42.65	43.20
80	43.74	44.29	44.84	45.38	45.93	46.48	47.03	47.57	48.12	48.67
90	49.21	49.76	50.31	50.85	51.40	51.95	52.49	53.04	53.59	54.13

Example: 982 m = 492.13 + 44.84 fathoms = 536.97 fathoms

Table 9 MILES TO KILOMETRES

1 mile = 1.609 344 km (exactly)
All values in Table 9(a) are exact.

(a)

miles	0	1000	2000	3000	4000	5000	6000	7000	8000	9000
0	—	1609.344	3218.688	4828.032	6437.376	8046.720	9656.064	11265.408	12874.752	14484.096
10000	16093.440	17702.784	19312.128	20921.472	22530.816	24140.160	25749.504	27358.848	28968.192	30577.536
20000	32186.880	33796.224	35405.568	37014.912	38624.256	40233.600	41842.944	43452.288	45061.632	46670.976

(b)

miles	0	1	2	3	4	5	6	7	8	9
0	—	1.609	3.219	4.828	6.437	8.047	9.656	11.265	12.875	14.484
10	16.093	17.703	19.312	20.921	22.531	24.140	25.750	27.359	28.968	30.578
20	32.187	33.796	35.406	37.015	38.624	40.234	41.843	43.452	45.062	46.671
30	48.280	49.890	51.499	53.108	54.718	56.327	57.936	59.546	61.155	62.764
40	64.374	65.983	67.592	69.202	70.811	72.420	74.030	75.639	77.249	78.858
50	80.467	82.077	83.686	85.295	86.905	88.514	90.123	91.733	93.342	94.951
60	96.561	98.170	99.779	101.389	102.998	104.607	106.217	107.826	109.435	111.045
70	112.654	114.263	115.873	117.482	119.091	120.701	122.310	123.919	125.529	127.138
80	128.748	130.357	131.966	133.576	135.185	136.794	138.404	140.013	141.622	143.232
90	144.841	146.450	148.060	149.669	151.278	152.888	154.497	156.106	157.716	159.325
100	160.934	162.544	164.153	165.762	167.372	168.981	170.590	172.200	173.809	175.418
110	177.028	178.637	180.247	181.856	183.465	185.075	186.684	188.293	189.903	191.512
120	193.121	194.731	196.340	197.949	199.559	201.168	202.777	204.387	205.996	207.605
130	209.215	210.824	212.433	214.043	215.652	217.261	218.871	220.480	222.089	223.699
140	225.308	226.918	228.527	230.136	231.746	233.355	234.964	236.574	238.183	239.792
150	241.402	243.011	244.620	246.230	247.839	249.448	251.058	252.667	254.276	255.886
160	257.495	259.104	260.714	262.323	263.932	265.542	267.151	268.760	270.370	271.979
170	273.588	275.198	276.807	278.417	280.026	281.635	283.245	284.854	286.463	288.073
180	289.682	291.291	292.901	294.510	296.119	297.729	299.338	300.947	302.557	304.166
190	305.775	307.385	308.994	310.603	312.213	313.822	315.431	317.041	318.650	320.259
200	321.869	323.478	325.087	326.697	328.306	329.916	331.525	333.134	334.744	336.353
210	337.962	339.572	341.181	342.790	344.400	346.009	347.618	349.228	350.837	352.446
220	354.056	355.665	357.274	358.884	360.493	362.102	363.712	365.321	366.930	368.540
230	370.149	371.758	373.368	374.977	376.586	378.196	379.805	381.415	383.024	384.633
240	386.243	387.852	389.461	391.071	392.680	394.289	395.899	397.508	399.117	400.727
250	402.336	403.945	405.555	407.164	408.773	410.383	411.992	413.601	415.211	416.820
260	418.429	420.039	421.648	423.257	424.867	426.476	428.086	429.695	431.304	432.914
270	434.523	436.132	437.742	439.351	440.960	442.570	444.179	445.788	447.398	449.007
280	450.616	452.226	453.835	455.444	457.054	458.663	460.272	461.882	463.491	465.100
290	466.710	468.319	469.928	471.538	473.147	474.756	476.366	477.975	479.585	481.194
300	482.803	484.413	486.022	487.631	489.241	490.850	492.459	494.069	495.678	497.287
310	498.897	500.506	502.115	503.725	505.334	506.943	508.553	510.162	511.771	513.381
320	514.990	516.599	518.209	519.818	521.427	523.037	524.646	526.255	527.865	529.474
330	531.084	532.693	534.302	535.912	537.521	539.130	540.740	542.349	543.958	545.568
340	547.177	548.786	550.396	552.005	553.614	555.224	556.833	558.442	560.052	561.661
350	563.270	564.880	566.489	568.098	569.708	571.317	572.926	574.536	576.145	577.754
360	579.364	580.973	582.583	584.192	585.801	587.411	589.020	590.629	592.239	593.848
370	595.457	597.067	598.676	600.285	601.895	603.504	605.113	606.723	608.332	609.941
380	611.551	613.160	614.769	616.379	617.988	619.597	621.207	622.816	624.425	626.035
390	627.644	629.254	630.863	632.472	634.082	635.691	637.300	638.910	640.519	642.128

miles	0	1	2	3	4	5	6	7	8	9
400	643.738	645.347	646.956	648.566	650.175	651.784	653.394	655.003	656.612	658.222
410	659.831	661.440	663.050	664.659	666.268	667.878	669.487	671.096	672.706	674.315
420	675.924	677.534	679.143	680.753	682.362	683.971	685.581	687.190	688.799	690.409
430	692.018	693.627	695.237	696.846	698.455	700.065	701.674	703.283	704.893	706.502
440	708.111	709.721	711.330	712.939	714.549	716.158	717.767	719.377	720.986	722.595
450	724.205	725.814	727.423	729.033	730.642	732.252	733.861	735.470	737.080	738.689
460	740.298	741.908	743.517	745.126	746.736	748.345	749.954	751.564	753.173	754.782
470	756.392	758.001	759.610	761.220	762.829	764.438	766.048	767.657	769.266	770.876
480	772.485	774.094	775.704	777.313	778.922	780.532	782.141	783.751	785.360	786.969
490	788.579	790.188	791.797	793.407	795.016	796.625	798.235	799.844	801.453	803.063
500	804.672	806.281	807.891	809.500	811.109	812.719	814.328	815.937	817.547	819.156
510	820.765	822.375	823.984	825.593	827.203	828.812	830.422	832.031	833.640	835.250
520	836.859	838.468	840.078	841.687	843.296	844.906	846.515	848.124	849.734	851.343
530	852.952	854.562	856.171	857.780	859.390	860.999	862.608	864.218	865.827	867.436
540	869.046	870.655	872.264	873.874	875.483	877.092	878.702	880.311	881.921	883.530
550	885.139	886.749	888.358	889.967	891.577	893.186	894.795	896.405	898.014	899.623
560	901.233	902.842	904.451	906.061	907.670	909.279	910.889	912.498	914.107	915.717
570	917.326	918.935	920.545	922.154	923.763	925.373	926.982	928.591	930.201	931.810
580	933.420	935.029	936.638	938.248	939.857	941.466	943.076	944.685	946.294	947.904
590	949.513	951.122	952.732	954.341	955.950	957.560	959.169	960.778	962.388	963.997
600	965.606	967.216	968.825	970.434	972.044	973.653	975.262	976.872	978.481	980.090
610	981.700	983.309	984.919	986.528	988.137	989.747	991.356	992.965	994.575	996.184
620	997.793	999.403	1001.012	1002.621	1004.231	1005.840	1007.449	1009.059	1010.668	1012.277
630	1013.887	1015.496	1017.105	1018.715	1020.324	1021.933	1023.543	1025.152	1026.761	1028.371
640	1029.980	1031.590	1033.199	1034.808	1036.418	1038.027	1039.636	1041.246	1042.855	1044.464
650	1046.074	1047.683	1049.292	1050.902	1052.511	1054.120	1055.730	1057.339	1058.948	1060.558
660	1062.167	1063.776	1065.386	1066.995	1068.604	1070.214	1071.823	1073.432	1075.042	1076.651
670	1078.260	1079.870	1081.479	1083.089	1084.698	1086.307	1087.917	1089.526	1091.135	1092.745
680	1094.354	1095.963	1097.573	1099.182	1100.791	1102.401	1104.010	1105.619	1107.229	1108.838
690	1110.447	1112.057	1113.666	1115.275	1116.885	1118.494	1120.103	1121.713	1123.322	1124.931
700	1126.541	1128.150	1129.759	1131.369	1132.978	1134.588	1136.197	1137.806	1139.416	1141.025
710	1142.634	1144.244	1145.853	1147.462	1149.072	1150.681	1152.290	1153.900	1155.509	1157.118
720	1158.728	1160.337	1161.946	1163.556	1165.165	1166.774	1168.384	1169.993	1171.602	1173.212
730	1174.821	1176.430	1178.040	1179.649	1181.258	1182.868	1184.477	1186.087	1187.696	1189.305
740	1190.915	1192.524	1194.133	1195.743	1197.352	1198.961	1200.571	1202.180	1203.789	1205.399
750	1207.008	1208.617	1210.227	1211.836	1213.445	1215.055	1216.664	1218.273	1219.883	1221.492
760	1223.101	1224.711	1226.320	1227.929	1229.539	1231.148	1232.758	1234.367	1235.976	1237.586
770	1239.195	1240.804	1242.414	1244.023	1245.632	1247.242	1248.851	1250.460	1252.070	1253.679
780	1255.288	1256.898	1258.507	1260.116	1261.726	1263.335	1264.944	1266.554	1268.163	1269.772
790	1271.382	1272.991	1274.600	1276.210	1277.819	1279.428	1281.038	1282.647	1284.257	1285.866
800	1287.475	1289.085	1290.694	1292.303	1293.913	1295.522	1297.131	1298.741	1300.350	1301.959
810	1303.569	1305.178	1306.787	1308.397	1310.006	1311.615	1313.225	1314.834	1316.443	1318.053
820	1319.662	1321.271	1322.881	1324.490	1326.099	1327.709	1329.318	1330.927	1332.537	1334.146
830	1335.756	1337.365	1338.974	1340.584	1342.193	1343.802	1345.412	1347.021	1348.630	1350.240
840	1351.849	1353.458	1355.068	1356.677	1358.286	1359.896	1361.505	1363.114	1364.724	1366.333

miles	0	1	2	3	4	5	6	7	8	9
850	1367.942	1369.552	1371.161	1372.770	1374.380	1375.989	1377.598	1379.208	1380.817	1382.426
860	1384.036	1385.645	1387.255	1388.864	1390.473	1392.083	1393.692	1395.301	1396.911	1398.520
870	1400.129	1401.739	1403.348	1404.957	1406.567	1408.176	1409.785	1411.395	1413.004	1414.613
880	1416.223	1417.832	1419.441	1421.051	1422.660	1424.269	1425.879	1427.488	1429.097	1430.707
890	1432.316	1433.926	1435.535	1437.144	1438.754	1440.363	1441.972	1443.582	1445.191	1446.800
900	1448.410	1450.019	1451.628	1453.238	1454.847	1456.456	1458.066	1459.675	1461.284	1462.894
910	1464.503	1466.112	1467.722	1469.331	1470.940	1472.550	1474.159	1475.768	1477.378	1478.987
920	1480.596	1482.206	1483.815	1485.425	1487.034	1488.643	1490.253	1491.862	1493.471	1495.081
930	1496.690	1498.299	1499.909	1501.518	1503.127	1504.737	1506.346	1507.955	1509.565	1511.174
940	1512.783	1514.393	1516.002	1517.611	1519.221	1520.830	1522.439	1524.049	1525.658	1527.267
950	1528.877	1530.486	1532.095	1533.705	1535.314	1536.924	1538.533	1540.142	1541.752	1543.361
960	1544.970	1546.580	1548.189	1549.798	1551.408	1553.017	1554.626	1556.236	1557.845	1559.454
970	1561.064	1562.673	1564.282	1565.892	1567.501	1569.110	1570.720	1572.329	1573.938	1575.548
980	1577.157	1578.766	1580.376	1581.985	1583.594	1585.204	1586.813	1588.423	1590.032	1591.641
990	1593.251	1594.860	1596.469	1598.079	1599.688	1601.297	1602.907	1604.516	1606.125	1607.735

Example: 24902 miles = 38624.256 + 1451.628 km = 40075.884 km
Note: The above table can also be used for converting miles per hour to kilometres per hour.
Examples: 30 miles h^{-1} = 48 km h^{-1} (approx.)
40 miles h^{-1} — 64 km h^{-1} (approx.)
70 miles h^{-1} = 113 km h^{-1} (approx.)

Table 10 KILOMETRES TO MILES

(a)

1 km = 0.621 371 192 mile

kilo-metres	0	1000	2000	3000	4000	5000	6000	7000	8000	9000
0	—	621.371	1242.742	1864.114	2485.485	3106.856	3728.227	4349.598	4970.970	5592.341
10000	6213.712	6835.083	7456.454	8077.825	8699.197	9320.568	9941.939	10563.310	11184.681	11806.053
20000	12427.424	13048.795	13670.166	14291.537	14912.909	15534.280	16155.651	16777.022	17398.393	18019.765

(b)

kilometres	0	1	2	3	4	5	6	7	8	9
0	—	0.621	1.243	1.864	2.485	3.107	3.728	4.350	4.971	5.592
10	6.214	6.835	7.456	8.078	8.699	9.321	9.942	10.563	11.185	11.806
20	12.427	13.049	13.670	14.292	14.913	15.534	16.156	16.777	17.398	18.020
30	18.641	19.263	19.884	20.505	21.127	21.748	22.369	22.991	23.612	24.233
40	24.855	25.476	26.098	26.719	27.340	27.962	28.583	29.204	29.826	30.447
50	31.069	31.690	32.311	32.933	33.554	34.175	34.797	35.418	36.040	36.661
60	37.282	37.904	38.525	39.146	39.768	40.389	41.010	41.632	42.253	42.875
70	43.496	44.117	44.739	45.360	45.981	46.603	47.224	47.846	48.467	49.088
80	49.710	50.331	50.952	51.574	52.195	52.817	53.438	54.059	54.681	55.302
90	55.923	56.545	57.166	57.788	58.409	59.030	59.652	60.273	60.894	61.516

km to miles

kilometres	0	1	2	3	4	5	6	7	8	9
100	62.137	62.758	63.380	64.001	64.623	65.244	65.865	66.487	67.108	67.729
110	68.351	68.972	69.594	70.215	70.836	71.458	72.079	72.700	73.322	73.943
120	74.565	75.186	75.807	76.429	77.050	77.671	78.293	78.914	79.536	80.157
130	80.778	81.400	82.021	82.642	83.264	83.885	84.506	85.128	85.749	86.371
140	86.992	87.613	88.235	88.856	89.477	90.099	90.720	91.342	91.963	92.584
150	93.206	93.827	94.448	95.070	95.691	96.313	96.934	97.555	98.177	98.798
160	99.419	100.041	100.662	101.284	101.905	102.526	103.148	103.769	104.390	105.012
170	105.633	106.254	106.876	107.497	108.119	108.740	109.361	109.983	110.604	111.225
180	111.847	112.468	113.090	113.711	114.332	114.954	115.575	116.196	116.818	117.439
190	118.061	118.682	119.303	119.925	120.546	121.167	121.789	122.410	123.031	123.653
200	124.274	124.896	125.517	126.138	126.760	127.381	128.002	128.624	129.245	129.867
210	130.488	131.109	131.731	132.352	132.973	133.595	134.216	134.838	135.459	136.080
220	136.702	137.323	137.944	138.566	139.187	139.809	140.430	141.051	141.673	142.294
230	142.915	143.537	144.158	144.779	145.401	146.022	146.644	147.265	147.886	148.508
240	149.129	149.750	150.372	150.993	151.615	152.236	152.857	153.479	154.100	154.721
250	155.343	155.964	156.586	157.207	157.828	158.450	159.071	159.692	160.314	160.935
260	161.557	162.178	162.799	163.421	164.042	164.663	165.285	165.906	166.527	167.149
270	167.770	168.392	169.013	169.634	170.256	170.877	171.498	172.120	172.741	173.363
280	173.984	174.605	175.227	175.848	176.469	177.091	177.712	178.334	178.955	179.576
290	180.198	180.819	181.440	182.062	182.683	183.305	183.926	184.547	185.169	185.790
300	186.411	187.033	187.654	188.275	188.897	189.518	190.140	190.761	191.382	192.004
310	192.625	193.246	193.868	194.489	195.111	195.732	196.353	196.975	197.596	198.217
320	198.839	199.460	200.082	200.703	201.324	201.946	202.567	203.188	203.810	204.431
330	205.052	205.674	206.295	206.917	207.538	208.159	208.781	209.402	210.023	210.645
340	211.266	211.888	212.509	213.130	213.752	214.373	214.994	215.616	216.237	216.859
350	217.480	218.101	218.723	219.344	219.965	220.587	221.208	221.830	222.451	223.072
360	223.694	224.315	224.936	225.558	226.179	226.800	227.422	228.043	228.665	229.286
370	229.907	230.529	231.150	231.771	232.393	233.014	233.636	234.257	234.878	235.500
380	236.121	236.742	237.364	237.985	238.607	239.228	239.849	240.471	241.092	241.713
390	242.335	242.956	243.578	244.199	244.820	245.442	246.063	246.684	247.306	247.927
400	248.548	249.170	249.791	250.413	251.034	251.655	252.277	252.898	253.519	254.141
410	254.762	255.384	256.005	256.626	257.248	257.869	258.490	259.112	259.733	260.355
420	260.976	261.597	262.219	262.840	263.461	264.083	264.704	265.325	265.947	266.568
430	267.190	267.811	268.432	269.054	269.675	270.296	270.918	271.539	272.161	272.782
440	273.403	274.025	274.646	275.267	275.889	276.510	277.132	277.753	278.374	278.996
450	279.617	280.238	280.860	281.481	282.103	282.724	283.345	283.967	284.588	285.209
460	285.831	286.452	287.073	287.695	288.316	288.938	289.559	290.180	290.802	291.423
470	292.044	292.666	293.287	293.909	294.530	295.151	295.773	296.394	297.015	297.637
480	298.258	298.880	299.501	300.122	300.744	301.365	301.986	302.608	303.229	303.851
490	304.472	305.093	305.715	306.336	306.957	307.579	308.200	308.821	309.443	310.064
500	310.686	311.307	311.928	312.550	313.171	313.792	314.414	315.035	315.657	316.278
510	316.899	317.521	318.142	318.763	319.385	320.006	320.628	321.249	321.870	322.492
520	323.113	323.734	324.356	324.977	325.599	326.220	326.841	327.463	328.084	328.705
530	329.327	329.948	330.569	331.191	331.812	332.434	333.055	333.676	334.298	334.919
540	335.540	336.162	336.783	337.405	338.026	338.647	339.269	339.890	340.511	341.133

kilometres	0	1	2	3	4	5	6	7	8	9
550	341.754	342.376	342.997	343.618	344.240	344.861	345.482	346.104	346.725	347.346
560	347.968	348.589	349.211	349.832	350.453	351.075	351.696	352.317	352.939	353.560
570	354.182	354.803	355.424	356.046	356.667	357.288	357.910	358.531	359.153	359.774
580	360.395	361.017	361.638	362.259	362.881	363.502	364.124	364.745	365.366	365.988
590	366.609	367.230	367.852	368.473	369.094	369.716	370.337	370.959	371.580	372.201
600	372.823	373.444	374.065	374.687	375.308	375.930	376.551	377.172	377.794	378.415
610	379.036	379.658	380.279	380.901	381.522	382.143	382.765	383.386	384.007	384.629
620	385.250	385.872	386.493	387.114	387.736	388.357	388.978	389.600	390.221	390.842
630	391.464	392.085	392.707	393.328	393.949	394.571	395.192	395.813	396.435	397.056
640	397.678	398.299	398.920	399.542	400.163	400.784	401.406	402.027	402.649	403.270
650	403.891	404.513	405.134	405.755	406.377	406.998	407.620	408.241	408.862	409.484
660	410.105	410.726	411.348	411.969	412.590	413.212	413.833	414.455	415.076	415.697
670	416.319	416.940	417.561	418.183	418.804	419.426	420.047	420.668	421.290	421.911
680	422.532	423.154	423.775	424.397	425.018	425.639	426.261	426.882	427.503	428.125
690	428.746	429.367	429.989	430.610	431.232	431.853	432.474	433.096	433.717	434.338
700	434.960	435.581	436.203	436.824	437.445	438.067	438.688	439.309	439.931	440.552
710	441.174	441.795	442.416	443.038	443.659	444.280	444.902	445.523	446.145	446.766
720	447.387	448.009	448.630	449.251	449.873	450.494	451.115	451.737	452.358	452.980
730	453.601	454.222	454.844	455.465	456.086	456.708	457.329	457.951	458.572	459.193
740	459.815	460.436	461.057	461.679	462.300	462.922	463.543	464.164	464.786	465.407
750	466.028	466.650	467.271	467.893	468.514	469.135	469.757	470.378	470.999	471.621
760	472.242	472.863	473.485	474.106	474.728	475.349	475.970	476.592	477.213	477.834
770	478.456	479.077	479.699	480.320	480.941	481.563	482.184	482.805	483.427	484.048
780	484.670	485.291	485.912	486.534	487.155	487.776	488.398	489.019	489.640	490.262
790	490.883	491.505	492.126	492.747	493.369	493.990	494.611	495.233	495.854	496.476
800	497.097	497.718	498.340	498.961	499.582	500.204	500.825	501.447	502.068	502.689
810	503.311	503.932	504.553	505.175	505.796	506.418	507.039	507.660	508.282	508.903
820	509.524	510.146	510.767	511.388	512.010	512.631	513.253	513.874	514.495	515.117
830	515.738	516.359	516.981	517.602	518.224	518.845	519.466	520.088	520.709	521.330
840	521.952	522.573	523.195	523.816	524.437	525.059	525.680	526.301	526.923	527.544
850	528.166	528.787	529.408	530.030	530.651	531.272	531.894	532.515	533.136	533.758
860	534.379	535.001	535.622	536.243	536.865	537.486	538.107	538.729	539.350	539.972
870	540.593	541.214	541.836	542.457	543.078	543.700	544.321	544.943	545.564	546.185
880	546.807	547.428	548.049	548.671	549.292	549.914	550.535	551.156	551.778	552.399
890	553.020	553.642	554.263	554.884	555.506	556.127	556.749	557.370	557.991	558.613
900	559.234	559.855	560.477	561.098	561.720	562.341	562.962	563.584	564.205	564.826
910	565.448	566.069	566.691	567.312	567.933	568.555	569.176	569.797	570.419	571.040
920	571.661	572.283	572.904	573.526	574.147	574.768	575.390	576.011	576.632	577.254
930	577.875	578.497	579.118	579.739	580.361	580.982	581.603	582.225	582.846	583.468
940	584.089	584.710	585.332	585.953	586.574	587.196	587.817	588.439	589.060	589.681
950	590.303	590.924	591.545	592.167	592.788	593.409	594.031	594.652	595.274	595.895
960	596.516	597.138	597.759	598.380	599.002	599.623	600.245	600.866	601.487	602.109
970	602.730	603.351	603.973	604.594	605.216	605.837	606.458	607.080	607.701	608.322
980	608.944	609.565	610.187	610.808	611.429	612.051	612.672	613.293	613.915	614.536
990	615.157	615.779	616.400	617.022	617.643	618.264	618.886	619.507	620.128	620.750

Example: 13455 km = 8077.825 + 282.724 miles = 8360.549 miles
See note at end of Table 9, page 23.

Table 11 INTERNATIONAL NAUTICAL MILES TO KILOMETRES

[n miles to km]   1 n mile = 1.852 km (exactly)
All values in Tables 11 (a) and 11 (b) are exact

(a)

nautical miles	0	100	200	300	400	500	600	700	800	900
0	—	185.2	370.4	555.6	740.8	926.0	1111.2	1296.4	1481.6	1666.8
1000	1852.0	2037.2	2222.4	2407.6	2592.8	2778.0	2963.2	3148.4	3333.6	3518.8

(b)

nautical miles	0	1	2	3	4	5	6	7	8	9
0	—	1.852	3.704	5.556	7.408	9.260	11.112	12.964	14.816	16.668
10	18.520	20.372	22.224	24.076	25.928	27.780	29.632	31.484	33.336	35.188
20	37.040	38.892	40.744	42.596	44.448	46.300	48.152	50.004	51.856	53.708
30	55.560	57.412	59.264	61.116	62.968	64.820	66.672	68.524	70.376	72.228
40	74.080	75.932	77.784	79.636	81.488	83.340	85.192	87.044	88.896	90.748
50	92.600	94.452	96.304	98.156	100.008	101.860	103.712	105.564	107.416	109.268
60	111.120	112.972	114.824	116.676	118.528	120.380	122.232	124.084	125.936	127.788
70	129.640	131.492	133.344	135.196	137.048	138.900	140.752	142.604	144.456	146.308
80	148.160	150.012	151.864	153.716	155.568	157.420	159.272	161.124	162.976	164.828
90	166.680	168.532	170.384	172.236	174.088	175.940	177.792	179.644	181.496	183.348

Example: 756 n miles = 1296.4 + 103.712 km = 1400.112 km

Table 12 KILOMETRES TO INTERNATIONAL NAUTICAL MILES

[km to n miles]   1 km = 0.539 956 8 n mile

(a)

kilometres	0	100	200	300	400	500	600	700	800	900
0	—	53.996	107.991	161.987	215.983	269.978	323.974	377.970	431.965	485.961
1000	539.957	593.952	647.948	701.944	755.940	809.935	863.931	917.927	971.922	1025.918

(b)

kilometres	0	1	2	3	4	5	6	7	8	9
0	—	0.540	1.080	1.620	2.160	2.700	3.240	3.780	4.320	4.860
10	5.400	5.940	6.479	7.019	7.559	8.099	8.639	9.179	9.719	10.259
20	10.799	11.339	11.879	12.419	12.959	13.499	14.039	14.579	15.119	15.659
30	16.199	16.739	17.279	17.819	18.359	18.898	19.438	19.978	20.518	21.058
40	21.598	22.138	22.678	23.218	23.758	24.298	24.838	25.378	25.918	26.458
50	26.998	27.538	28.078	28.618	29.158	29.698	30.238	30.778	31.317	31.857
60	32.397	32.937	33.477	34.017	34.557	35.097	35.637	36.177	36.717	37.257
70	37.797	38.337	38.877	39.417	39.957	40.497	41.037	41.577	42.117	42.657
80	43.197	43.737	44.276	44.816	45.356	45.896	46.436	46.976	47.516	48.056
90	48.596	49.136	49.676	50.216	50.756	51.296	51.836	52.376	52.916	53.456

Example: 1654 km = 863.931 + 29.158 n miles = 893.089 n miles
Note: Since·one *international nautical mile per hour* is the same as one *international knot* (*kn*), Table 11 can also be used for converting international knots to kilometres per hour, and Table 12 for converting kilometres per hour to international knots.
Example: 15 kn = 27.78 km h^{-1}

Table 13 INTERNATIONAL NAUTICAL MILES TO MILES

1 n mile = 1.150 779 45 miles

(a)

nautical miles	0	100	200	300	400	500	600	700	800	900
0	—	115.078	230.156	345.234	460.312	575.390	690.468	805.546	920.624	1035.702
1000	1150.779	1265.857	1380.935	1496.013	1611.091	1726.169	1841.247	1956.325	2071.403	2186.481

(b)

nautical miles	0	1	2	3	4	5	6	7	8	9
0	—	1.151	2.302	3.452	4.603	5.754	6.905	8.055	9.206	10.357
10	11.508	12.659	13.809	14.960	16.111	17.262	18.412	19.563	20.714	21.865
20	23.016	24.166	25.317	26.468	27.619	28.769	29.920	31.071	32.222	33.373
30	34.523	35.674	36.825	37.976	39.127	40.277	41.428	42.579	43.730	44.880
40	46.031	47.182	48.333	49.484	50.634	51.785	52.936	54.087	55.237	56.388
50	57.539	58.690	59.841	60.991	62.142	63.293	64.444	65.594	66.745	67.896
60	69.047	70.198	71.348	72.499	73.650	74.801	75.951	77.102	78.253	79.404
70	80.555	81.705	82.856	84.007	85.158	86.308	87.459	88.610	89.761	90.912
80	92.062	93.213	94.364	95.515	96.665	97.816	98.967	100.118	101.269	102.419
90	103.570	104.721	105.872	107.022	108.173	109.324	110.475	111.626	112.776	113.927

Example: 692 n miles = 690.468 + 105.872 miles = 796.340 miles

Table 14 MILES TO INTERNATIONAL NAUTICAL MILES

1 mile = 0.868 976 242 n mile

(a)

miles	0	100	200	300	400	500	600	700	800	900
0	—	86.898	173.795	260.693	347.590	434.488	521.386	608.283	695.181	782.079
1000	868.976	955.874	1042.771	1129.669	1216.567	1303.464	1390.362	1477.260	1564.157	1651.055

(b)

miles	0	1	2	3	4	5	6	7	8	9
0	—	0.869	1.738	2.607	3.476	4.345	5.214	6.083	6.952	7.821
10	8.690	9.559	10.428	11.297	12.166	13.035	13.904	14.773	15.642	16.511
20	17.380	18.249	19.117	19.986	20.855	21.724	22.593	23.462	24.331	25.200
30	26.069	26.938	27.807	28.676	29.545	30.414	31.283	32.152	33.021	33.890
40	34.759	35.628	36.497	37.366	38.235	39.104	39.973	40.842	41.711	42.580
50	43.449	44.318	45.187	46.056	46.925	47.794	48.663	49.532	50.401	51.270
60	52.139	53.008	53.877	54.746	55.614	56.483	57.352	58.221	59.090	59.959
70	60.828	61.697	62.566	63.435	64.304	65.173	66.042	66.911	67.780	68.649
80	69.518	70.387	71.256	72.125	72.994	73.863	74.732	75.601	76.470	77.339
90	78.208	79.077	79.946	80.815	81.684	82.553	83.422	84.291	85.160	86.029

Example: 1123 miles = 955.874 + 19.986 n miles = 975.860 n miles

Note: (i) Conversion factors:

1 UK nautical mile = 1.000 64 n mile

1 n mile = 0.999 361 UK nautical mile

(ii) The United Kingdom adopted the international nautical mile in place of the UK nautical mile in 1970.

Table 15 SQUARE INCHES TO SQUARE CENTIMETRES

[in² to cm²]   1 in² = 6.451 6 cm² (exactly)
All values in Table 15(a) are exact

(a)

square inches	0	100	200	300	400	500	600	700	800	900
0	—	645.16	1290.32	1935.48	2580.64	3225.80	3870.96	4516.12	5161.28	5806.44
1000	6451.60	7096.76	7741.92	8387.08	9032.24	9677.40	10322.56	10967.72	11612.88	12258.04

(b)

square inches	0	1	2	3	4	5	6	7	8	9
0	—	06.452	12.903	19.355	25.806	32.258	38.710	45.161	51.613	58.064
10	64.516	70.968	77.419	83.871	90.322	96.774	103.226	109.677	116.129	122.580
20	129.032	135.484	141.935	148.387	154.838	161.290	167.742	174.193	180.645	187.096
30	193.548	200.000	206.451	212.903	219.354	225.806	232.258	238.709	245.161	251.612
40	258.064	264.516	270.967	277.419	283.870	290.322	296.774	303.225	309.677	316.128
50	322.580	329.032	335.483	341.935	348.386	354.838	361.290	367.741	374.193	380.644
60	387.096	393.548	399.999	406.451	412.902	419.354	425.806	432.257	438.709	445.160
70	451.612	458.064	464.515	470.967	477.418	483.870	490.322	496.773	503.225	509.676
80	516.128	522.580	529.031	535.483	541.934	548.386	554.838	561.289	567.741	574.192
90	580.644	587.096	593.547	599.999	606.450	612.902	619.354	625.805	632.257	638.708

Example: 468 in² = 2580.64 + 438.709 cm² = 3019.349 cm²
Note: In order to convert:
(i)   square inches to square millimetres, shift the decimal point in Table 15 two places to the right;
(ii) square inches to square metres, shift the decimal point in Table 15 four places to the left.

Table 16 SQUARE CENTIMETRES TO SQUARE INCHES

[cm² to in²]   1 cm² = 0.155 000 31 in²

(a)

square centi-metres	0	100	200	300	400	500	600	700	800	900
0	—	15.50003	31.00006	46.50009	62.00012	77.50016	93.00019	108.50022	124.00025	139.50028
1000	155.00031	170.50034	186.00037	201.50040	217.00043	232.50047	248.00050	263.50053	279.00056	294.50059

(b)

square centi-metres	0	1	2	3	4	5	6	7	8	9
0	—	0.15500	0.31000	0.46500	0.62000	0.77500	0.93000	1.08500	1.24000	1.39500
10	1.55000	1.70500	1.86000	2.01500	2.17000	2.32500	2.48000	2.63501	2.79001	2.94501
20	3.10001	3.25501	3.41001	3.56501	3.72001	3.87501	4.03001	4.18501	4.34001	4.49501
30	4.65001	4.80501	4.96001	5.11501	5.27001	5.42501	5.58001	5.73501	5.89001	6.04501
40	6.20001	6.35501	6.51001	6.66501	6.82001	6.97501	7.13001	7.28501	7.44001	7.59502
50	7.75002	7.90502	8.06002	8.21502	8.37002	8.52502	8.68002	8.83502	8.99002	9.14502
60	9.30002	9.45502	9.61002	9.76502	9.92002	10.07502	10.23002	10.38502	10.54002	10.69502
70	10.85002	11.00502	11.16002	11.31502	11.47002	11.62502	11.78002	11.93502	12.09002	12.24502
80	12.40002	12.55503	12.71003	12.86503	13.02003	13.17503	13.33003	13.48503	13.64003	13.79503
90	13.95003	14.10503	14.26003	14.41503	14.57003	14.72503	14.88003	15.03503	15.19003	15.34503

Example: 375 cm² = 46.50009 + 11.62502 in² = 58.12511 in²

Table 17 SQUARE FEET TO SQUARE METRES

[ft² to m²]   1 ft² = 0.092 903 04 m² (exactly)

(a)

square feet	0	100	200	300	400	500	600	700	800	900
0	—	9.29030	18.58061	27.87091	37.16122	46.45152	55.74182	65.03213	74.32243	83.61274
1000	92.90304	102.19334	111.48365	120.77395	130.06426	139.35456	148.64486	157.93517	167.22547	176.51578
2000	185.80608	195.09638	204.38669	213.67699	222.96730	232.25760	241.54790	250.83821	260.12851	269.41882

(b)

square feet	0	1	2	3	4	5	6	7	8	9
0	—	0.09290	0.18581	0.27871	0.37161	0.46452	0.55742	0.65032	0.74322	0.83613
10	0.92903	1.02193	1.11484	1.20774	1.30064	1.39355	1.48645	1.57935	1.67225	1.76516
20	1.85806	1.95096	2.04387	2.13677	2.22967	2.32258	2.41548	2.50838	2.60129	2.69419
30	2.78709	2.87999	2.97290	3.06580	3.15870	3.25161	3.34451	3.43741	3.53032	3.62322
40	3.71612	3.80902	3.90193	3.99483	4.08773	4.18064	4.27354	4.36644	4.45935	4.55225
50	4.64515	4.73806	4.83096	4.92386	5.01676	5.10967	5.20257	5.29547	5.38838	5.48128
60	5.57418	5.66709	5.75999	5.85289	5.94579	6.03870	6.13160	6.22450	6.31741	6.41031
70	6.50321	6.59612	6.68902	6.78192	6.87482	6.96773	7.06063	7.15353	7.24644	7.33934
80	7.43224	7.52515	7.61805	7.71095	7.80386	7.89676	7.98966	8.08256	8.17547	8.26837
90	8.36127	8.45418	8.54708	8.63998	8.73289	8.82579	8.91869	9.01159	9.10450	9.19740

Example: 2587 ft² = 232.25760 + 8.08256 m² = 240.34016 m²

Table 18 SQUARE METRES TO SQUARE FEET

[m² to ft²]   1 m² = 10.763 910 4 ft²

(a)

square metres	0	100	200	300	400	500	600	700	800	900
0	—	1076.391	2152.782	3229.173	4305.564	5381.955	6458.346	7534.737	8611.128	9687.519
1000	10763.910	11840.301	12916.692	13993.084	15069.475	16145.866	17222.257	18298.648	19375.039	20451.430
2000	21527.821	22604.212	23680.603	24756.994	25833.385	26909.776	27986.167	29062.558	30138.949	31215.340

(b)

square metres	0	1	2	3	4	5	6	7	8	9
0	—	10.764	21.528	32.292	43.056	53.820	64.583	75.347	86.111	96.875
10	107.639	118.403	129.167	139.931	150.695	161.459	172.223	182.986	193.750	204.514
20	215.278	226.042	236.806	247.570	258.334	269.098	279.862	290.626	301.389	312.153
30	322.917	333.681	344.445	355.209	365.973	376.737	387.501	398.265	409.029	419.793
40	430.556	441.320	452.084	462.848	473.612	484.376	495.140	505.904	516.668	527.432
50	538.196	548.959	559.723	570.487	581.251	592.015	602.779	613.543	624.307	635.071
60	645.835	656.599	667.362	678.126	688.890	699.654	710.418	721.182	731.946	742.710
70	753.474	764.238	775.002	785.765	796.529	807.293	818.057	828.821	839.585	850.349
80	861.113	871.877	882.641	893.405	904.168	914.932	925.696	936.460	947.224	957.988
90	968.752	979.516	990.280	1001.044	1011.808	1022.571	1033.335	1044.099	1054.863	1065.627

Example: 1863 m² = 19375.039 + 678.126 ft² = 20053.165 ft²

Note: Table 17 can also be used for converting square feet per second to square metres per second, and Table 18 for converting square metres per second to square feet per second.

Table 19 SQUARE YARDS TO SQUARE METRES

[yd² to m²]   1 yd² = 0.836 127 36 m² (exactly)

(a)

square yards	0	100	200	300	400	500	600	700	800	900
0	—	83.6127	167.2255	250.8382	334.4509	418.0637	501.6764	585.2892	668.9019	752.5146
1000	836.1274	919.7401	1003.3528	1086.9656	1170.5783	1254.1910	1337.8038	1421.4165	1505.0292	1588.6420
2000	1672.2547	1755.8675	1839.4802	1923.0929	2006.7057	2090.3184	2173.9311	2257.5439	2341.1566	2424.7693

(b)

square yards	0	1	2	3	4	5	6	7	8	9
0	—	0.8361	1.6723	2.5084	3.3445	4.1806	5.0168	5.8529	6.6890	7.5251
10	8.3613	9.1974	10.0335	10.8697	11.7058	12.5419	13.3780	14.2142	15.0503	15.8864
20	16.7225	17.5587	18.3948	19.2309	20.0671	20.9032	21.7393	22.5754	23.4116	24.2477
30	25.0838	25.9199	26.7561	27.5922	28.4283	29.2645	30.1006	30.9367	31.7728	32.6090
40	33.4451	34.2812	35.1173	35.9535	36.7896	37.6257	38.4619	39.2980	40.1341	40.9702
50	41.8064	42.6425	43.4786	44.3148	45.1509	45.9870	46.8231	47.6593	48.4954	49.3315
60	50.1676	51.0038	51.8399	52.6760	53.5122	54.3483	55.1844	56.0205	56.8567	57.6928
70	58.5289	59.3650	60.2012	61.0373	61.8734	62.7096	63.5457	64.3818	65.2179	66.0541
80	66.8902	67.7263	68.5624	69.3986	70.2347	71.0708	71.9070	72.7431	73.5792	74.4153
90	75.2515	76.0876	76.9237	77.7598	78.5960	79.4321	80.2682	81.1044	81.9405	82.7766

Example: 2987 yd² = 2424.7693 + 72.7431 m² = 2497.5124 m²

Table 20 SQUARE METRES TO SQUARE YARDS

[m² to yd²]   1 m² = 1.195 990 046 3 yd²

(a)

square metres	0	100	200	300	400	500	600	700	800	900
0	—	119.5990	239.1980	358.7970	478.3960	597.9950	717.5940	837.1930	956.7920	1076.3910
1000	1195.9900	1315.5891	1435.1881	1554.7871	1674.3861	1793.9851	1913.5841	2033.1831	2152.7821	2272.3811
2000	2391.9801	2511.5791	2631.1781	2750.7771	2870.3761	2989.9751	3109.5741	3229.1731	3348.7721	3468.3711

(b)

square metres	0	1	2	3	4	5	6	7	8	9
0	—	1.1960	2.3920	3.5880	4.7840	5.9800	7.1759	8.3719	9.5679	10.7639
10	11.9599	13.1559	14.3519	15.5479	16.7439	17.9399	19.1358	20.3318	21.5278	22.7238
20	23.9198	25.1158	26.3118	27.5078	28.7038	29.8998	31.0957	32.2917	33.4877	34.6837
30	35.8797	37.0757	38.2717	39.4677	40.6637	41.8597	43.0556	44.2516	45.4476	46.6436
40	47.8396	49.0356	50.2316	51.4276	52.6236	53.8196	55.0155	56.2115	57.4075	58.6035
50	59.7995	60.9955	62.1915	63.3875	64.5835	65.7795	66.9754	68.1714	69.3674	70.5634
60	71.7594	72.9554	74.1514	75.3474	76.5434	77.7394	78.9353	80.1313	81.3273	82.5233
70	83.7193	84.9153	86.1113	87.3073	88.5033	89.6993	90.8952	92.0912	93.2872	94.4832
80	95.6792	96.8752	98.0712	99.2672	100.4632	101.6592	102.8551	104.0511	105.2471	106.4431
90	107.6391	108.8351	110.0311	111.2271	112.4231	113.6191	114.8150	116.0110	117.2070	118.4030

Example: 589 m² = 597.9950 + 106.4431 yd² = 704.4381 yd²

Table 21 ACRES TO HECTARES

Special Note regarding the conversion of Acres, Roods and Poles to Hectares
Where the area is given in acres, roods and poles.
1. *First* convert the area into acres and decimals of an acre.
  1 rood = ¼ acre = 0.25 acre
  1 pole = 1/40 rood = 0.00625 acre (use table below where necessary)
POLES TO DECIMALS OF AN ACRE [All values in this table are exact]

poles	0	1	2	3	4	5	6	7	8	9
0	—	0.00625	0.01250	0.01875	0.02500	0.03125	0.03750	0.04375	0.05000	0.05625
10	0.06250	0.06875	0.07500	0.08125	0.08750	0.09375	0.10000	0.10625	0.11250	0.11875
20	0.12500	0.13125	0.13750	0.14375	0.15000	0.15625	0.16250	0.16875	0.17500	0.18125
30	0.18750	0.19375	0.20000	0.20625	0.21250	0.21875	0.22500	0.23125	0.23750	0.24375

2. *Then* by the use of tables (a), (b) and (c) which follow, convert into hectares.
Example: 4 acres, 3 roods and 3 poles = 4.0 + 0.75 + 0.01875 acres = 4.76875 acres = 4.769 acres (correct to three decimal places) From tables (b) and (c) below, 4.769 acres = 1.6187 + 0.3112 ha = 1.9299 ha = 1.930 ha

(a) **ACRES TO HECTARES** [acres to ha] 1 acre = 0.404 685 64 ha

acres	0	100	200	300	400	500	600	700	800	900
0	—	40.4686	80.9371	121.4057	161.8743	202.3428	242.8114	283.2799	323.7485	364.2171
1000	404.6856	445.1542	485.6228	526.0913	566.5599	607.0285	647.4970	687.9656	728.4342	768.9027
2000	809.3713	849.8398	890.3084	930.7770	971.2455	1011.7141	1052.1827	1092.6512	1133.1198	1173.5884

(b)

acres	0	1	2	3	4	5	6	7	8	9
0	—	0.4047	0.8094	1.2141	1.6187	2.0234	2.4281	2.8328	3.2375	3.6422
10	4.0469	4.4515	4.8562	5.2609	5.6656	6.0703	6.4750	6.8797	7.2843	7.6890
20	8.0937	8.4984	8.9031	9.3078	9.7125	10.1171	10.5218	10.9265	11.3312	11.7359
30	12.1406	12.5453	12.9499	13.3546	13.7593	14.1640	14.5687	14.9734	15.3781	15.7827
40	16.1874	16.5921	16.9968	17.4015	17.8062	18.2109	18.6155	19.0202	19.4249	19.8296
50	20.2343	20.6390	21.0437	21.4483	21.8530	22.2577	22.6624	23.0671	23.4718	23.8765
60	24.2811	24.6858	25.0905	25.4952	25.8999	26.3046	26.7093	27.1139	27.5186	27.9233
70	28.3280	28.7327	29.1374	29.5421	29.9467	30.3514	30.7561	31.1608	31.5655	31.9702
80	32.3749	32.7795	33.1842	33.5889	33.9936	34.3983	34.8030	35.2077	35.6123	36.0170
90	36.4217	36.8264	37.2311	37.6358	38.0405	38.4451	38.8498	39.2545	39.6592	40.0639

(c)

acres	0.000	0.001	0.002	0.003	0.004	0.005	0.006	0.007	0.008	0.009
0.000	—	0.0004	0.0008	0.0012	0.0016	0.0020	0.0024	0.0028	0.0032	0.0036
0.010	0.0040	0.0045	0.0049	0.0053	0.0057	0.0061	0.0065	0.0069	0.0073	0.0077
0.020	0.0081	0.0085	0.0089	0.0093	0.0097	0.0101	0.0105	0.0109	0.0113	0.0117
0.030	0.0121	0.0125	0.0129	0.0134	0.0138	0.0142	0.0146	0.0150	0.0154	0.0158
0.040	0.0162	0.0166	0.0170	0.0174	0.0178	0.0182	0.0186	0.0190	0.0194	0.0198
0.050	0.0202	0.0206	0.0210	0.0214	0.0219	0.0223	0.0227	0.0231	0.0235	0.0239
0.060	0.0243	0.0247	0.0251	0.0255	0.0259	0.0263	0.0267	0.0271	0.0275	0.0279
0.070	0.0283	0.0287	0.0291	0.0295	0.0299	0.0304	0.0308	0.0312	0.0316	0.0320
0.080	0.0324	0.0328	0.0332	0.0336	0.0340	0.0344	0.0348	0.0352	0.0356	0.0360
0.090	0.0364	0.0368	0.0372	0.0376	0.0380	0.0384	0.0388	0.0393	0.0397	0.0401

acres to hectares

acres	0.000	0.001	0.002	0.003	0.004	0.005	0.006	0.007	0.008	0.009
0.100	0.0405	0.0409	0.0413	0.0417	0.0421	0.0425	0.0429	0.0433	0.0437	0.0441
0.110	0.0445	0.0449	0.0453	0.0457	0.0461	0.0465	0.0469	0.0473	0.0478	0.0482
0.120	0.0486	0.0490	0.0494	0.0498	0.0502	0.0506	0.0510	0.0514	0.0518	0.0522
0.130	0.0526	0.0530	0.0534	0.0538	0.0542	0.0546	0.0550	0.0554	0.0558	0.0563
0.140	0.0567	0.0571	0.0575	0.0579	0.0583	0.0587	0.0591	0.0595	0.0599	0.0603
0.150	0.0607	0.0611	0.0615	0.0619	0.0623	0.0627	0.0631	0.0635	0.0639	0.0643
0.160	0.0647	0.0652	0.0656	0.0660	0.0664	0.0668	0.0672	0.0676	0.0680	0.0684
0.170	0.0688	0.0692	0.0696	0.0700	0.0704	00708	0.0712	0.0716	0.0720	0.0724
0.180	0.0728	0.0732	0.0737	0.0741	0.0745	0.0749	0.0753	0.0757	0.0761	0.0765
0.190	0.0769	0.0773	0.0777	0.0781	0.0785	0.0789	0.0793	0.0797	0.0801	0.0805
0.200	0.0809	0.0813	0.0817	0.0822	0.0826	0.0830	0.0834	0.0838	0.0842	0.0846
0.210	0.0850	0.0854	0.0858	0.0862	0.0866	0.0870	0.0874	0.0878	0.0882	0.0886
0.220	0.0890	0.0894	0.0898	0.0902	0.0906	0.0911	0.0915	0.0919	0.0923	0.0927
0.230	0.0931	0.0935	0.0939	0.0943	0.0947	0.0951	0.0955	0.0959	0.0963	0.0967
0.240	0.0971	0.0975	0.0979	0.0983	0.0987	0.0991	0.0996	0.1000	0.1004	0.1008
0.250	0.1012	0.1016	0.1020	0.1024	0.1028	0.1032	0.1036	0.1040	0.1044	0.1048
0.260	0.1052	0.1056	0.1060	0.1064	0.1068	0.1072	0.1076	0.1081	0.1085	0.1089
0.270	0.1093	0.1097	0.1101	0.1105	0.1109	0.1113	0.1117	0.1121	0.1125	0.1129
0.280	0.1133	0.1137	0.1141	0.1145	0.1149	0.1153	0.1157	0.1161	0.1165	0.1170
0.290	0.1174	0.1178	0.1182	0.1186	0.1190	0.1194	0.1198	0.1202	0.1206	0.1210
0.300	0.1214	0.1218	0.1222	0.1226	0.1230	0.1234	0.1238	0.1242	0.1246	0.1250
0.310	0.1255	0.1259	0.1263	0.1267	0.1271	0.1275	0.1279	0.1283	0.1287	0.1291
0.320	0.1295	0.1299	0.1303	0.1307	0.1311	0.1315	0.1319	0.1323	0.1327	0.1331
0.330	0.1335	0.1340	0.1344	0.1348	0.1352	0.1356	0.1360	0.1364	0.1368	0.1372
0.340	0.1376	0.1380	0.1384	0.1388	0.1392	0.1396	0.1400	0.1404	0.1408	0.1412
0.350	0.1416	0.1420	0.1424	0.1429	0.1433	0.1437	0.1441	0.1445	0.1449	0.1453
0.360	0.1457	0.1461	0.1465	0.1469	0.1473	0.1477	0.1481	0.1485	0.1489	0.1493
0.370	0.1497	0.1501	0.1505	0.1509	0.1514	0.1518	0.1522	0.1526	0.1530	0.1534
0.380	0.1538	0.1542	0.1546	0.1550	0.1554	0.1558	0.1562	0.1566	0.1570	0.1574
0.390	0.1578	0.1582	0.1586	0.1590	0.1594	0.1599	0.1603	0.1607	0.1611	0.1615
0.400	0.1619	0.1623	0.1627	0.1631	0.1635	0.1639	0.1643	0.1647	0.1651	0.1655
0.410	0.1659	0.1663	0.1667	0.1671	0.1675	0.1679	0.1683	0.1688	0.1692	0.1696
0.420	0.1700	0.1704	0.1708	0.1712	0.1716	0.1720	0.1724	0.1728	0.1732	0.1736
0.430	0.1740	0.1744	0.1748	0.1752	0.1756	0.1760	0.1764	0.1768	0.1773	0.1777
0.440	0.1781	0.1785	0.1789	0.1793	0.1797	0.1801	0.1805	0.1809	0.1813	0.1817
0.450	0.1821	0.1825	0.1829	0.1833	0.1837	0.1841	0.1845	0.1849	0.1853	0.1858
0.460	0.1862	0.1866	0.1870	0.1874	0.1878	0.1882	0.1886	0.1890	0.1894	0.1898
0.470	0.1902	0.1906	0.1910	0.1914	0.1918	0.1922	0.1926	0.1930	0.1934	0.1938
0.480	0.1942	0.1947	0.1951	0.1955	0.1959	0.1963	0.1967	0.1971	0.1975	0.1979
0.490	0.1983	0.1987	0.1991	0.1995	0.1999	0.2003	0.2007	0.2011	0.2015	0.2019
0.500	0.2023	0.2027	0.2032	0.2036	0.2040	0.2044	0.2048	0.2052	0.2056	0.2060
0.510	0.2064	0.2068	0.2072	0.2076	0.2080	0.2084	0.2088	0.2092	0.2096	0.2100
0.520	0.2104	0.2108	0.2112	0.2117	0.2121	0.2125	0.2129	0.2133	0.2137	0.2141
0.530	0.2145	0.2149	0.2153	0.2157	0.2161	0.2165	0.2169	0.2173	0.2177	0.2181
0.540	0.2185	0.2189	0.2193	0.2197	0.2201	0.2206	0.2210	0.2214	0.2218	0.2222

acres	0.000	0.001	0.002	0.003	0.004	0.005	0.006	0.007	0.008	0.009
0.550	0.2226	0.2230	0.2234	0.2238	0.2242	0.2246	0.2250	0.2254	0.2258	0.2262
0.560	0.2266	0.2270	0.2274	0.2278	0.2282	0.2286	0.2291	0.2295	0.2299	0.2303
0.570	0.2307	0.2311	0.2315	0.2319	0.2323	0.2327	0.2331	0.2335	0.2339	0.2343
0.580	0.2347	0.2351	0.2355	0.2359	0.2363	0.2367	0.2371	0.2376	0.2380	0.2384
0.590	0.2388	0.2392	0.2396	0.2400	0.2404	0.2408	0.2412	0.2416	0.2420	0.2424
0.600	0.2428	0.2432	0.2436	0.2440	0.2444	0.2448	0.2452	0.2456	0.2460	0.2465
0.610	0.2469	0.2473	0.2477	0.2481	0.2485	0.2489	0.2493	0.2497	0.2501	0.2505
0.620	0.2509	0.2513	0.2517	0.2521	0.2525	0.2529	0.2533	0.2537	0.2541	0.2545
0.630	0.2550	0.2554	0.2558	0.2562	0.2566	0.2570	0.2574	0.2578	0.2582	0.2586
0.640	0.2590	0.2594	0.2598	0.2602	0.2606	0.2610	0.2614	0.2618	0.2622	0.2626
0.650	0.2630	0.2635	0.2639	0.2643	0.2647	0.2651	0.2655	0.2659	0.2663	0.2667
0.660	0.2671	0.2675	0.2679	0.2683	0.2687	0.2691	0.2695	0.2699	0.2703	0.2707
0.670	0.2711	0.2715	0.2719	0.2724	0.2728	0.2732	0.2736	0.2740	0.2744	0.2748
0.680	0.2752	0.2756	0.2760	0.2764	0.2768	0.2772	0.2776	0.2780	0.2784	0.2788
0.690	0.2792	0.2796	0.2800	0.2804	0.2809	0.2813	0.2817	0.2821	0.2825	0.2829
0.700	0.2833	0.2837	0.2841	0.2845	0.2849	0.2853	0.2857	0.2861	0.2865	0.2869
0.710	0.2873	0.2877	0.2881	0.2885	0.2889	0.2894	0.2898	0.2902	0.2906	0.2910
0.720	0.2914	0.2918	0.2922	0.2926	0.2930	0.2934	0.2938	0.2942	0.2946	0.2950
0.730	0.2954	0.2958	0.2962	0.2966	0.2970	0.2974	0.2978	0.2983	0.2987	0.2991
0.740	0.2995	0.2999	0.3003	0.3007	0.3011	0.3015	0.3019	0.3023	0.3027	0.3031
0.750	0.3035	0.3039	0.3043	0.3047	0.3051	0.3055	0.3059	0.3063	0.3068	0.3072
0.760	0.3076	0.3080	0.3084	0.3088	0.3092	0.3096	0.3100	0.3104	0.3108	0.3112
0.770	0.3116	0.3120	0.3124	0.3128	0.3132	0.3136	0.3140	0.3144	0.3148	0.3153
0.780	0.3157	0.3161	0.3165	0.3169	0.3173	0.3177	0.3181	0.3185	0.3189	0.3193
0.790	0.3197	0.3201	0.3205	0.3209	0.3213	0.3217	0.3221	0.3225	0.3229	0.3233
0.800	0.3237	0.3242	0.3246	0.3250	0.3254	0.3258	0.3262	0.3266	0.3270	0.3274
0.810	0.3278	0.3282	0.3286	0.3290	0.3294	0.3298	0.3302	0.3306	0.3310	0.3314
0.820	0.3318	0.3322	0.3327	0.3331	0.3335	0.3339	0.3343	0.3347	0.3351	0.3355
0.830	0.3359	0.3363	0.3367	0.3371	0.3375	0.3379	0.3383	0.3387	0.3391	0.3395
0.840	0.3399	0.3403	0.3407	0.3411	0.3416	0.3420	0.3424	0.3428	0.3432	0.3436
0.850	0.3440	0.3444	0.3448	0.3452	0.3456	0.3460	0.3464	0.3468	0.3472	0.3476
0.860	0.3480	0.3484	0.3488	0.3492	0.3496	0.3501	0.3505	0.3509	0.3513	0.3517
0.870	0.3521	0.3525	0.3529	0.3533	0.3537	0.3541	0.3545	0.3549	0.3553	0.3557
0.880	0.3561	0.3565	0.3569	0.3573	0.3577	0.3581	0.3586	0.3590	0.3594	0.3598
0.890	0.3602	0.3606	0.3610	0.3614	0.3618	0.3622	0.3626	0.3630	0.3634	0.3638
0.900	0.3642	0.3646	0.3650	0.3654	0.3658	0.3662	0.3666	0.3670	0.3675	0.3679
0.910	0.3683	0.3687	0.3691	0.3695	0.3699	0.3703	0.3707	0.3711	0.3715	0.3719
0.920	0.3723	0.3727	0.3731	0.3735	0.3739	0.3743	0.3747	0.3751	0.3755	0.3760
0.930	0.3764	0.3768	0.3772	0.3776	0.3780	0.3784	0.3788	0.3792	0.3796	0.3800
0.940	0.3804	0.3808	0.3812	0.3816	0.3820	0.3824	0.3828	0.3832	0.3836	0.3840
0.950	0.3845	0.3849	0.3853	0.3857	0.3861	0.3865	0.3869	0.3873	0.3877	0.3881
0.960	0.3885	0.3889	0.3893	0.3897	0.3901	0.3905	0.3909	0.3913	0.3917	0.3921
0.970	0.3925	0.3929	0.3934	0.3938	0.3942	0.3946	0.3950	0.3954	0.3958	0.3962
0.980	0.3966	0.3970	0.3974	0.3978	0.3982	0.3986	0.3990	03994	0.3998	0.4002
0.990	0.4006	0.4010	0.4014	0.4019	0.4023	0.4027	0.4031	0.4035	0.4039	0.4043

Example: 640 acres = 242.8114 + 16.1874 ha = 258.9988 ha

Note: Ordnance Survey maps (scale 1/2500). On new Ordnance Survey maps areas of fields, etc. are shown in hectares to three decimal places; on old Ordnance Survey maps they are shown in acres to three decimal places. To convert areas in acres shown on old maps to hectares, use Table 21 and round off to three places of decimals

Table 22 HECTARES TO ACRES
[ha to acres] 1 ha = 2.471 053 81 acres

(a)

hectares	0	100	200	300	400	500	600	700	800	900
0	—	247.105	494.211	741.316	988.422	1235.527	1482.632	1729.738	1976.843	2223.948
1000	2471.054	2718.159	2965.265	3212.370	3459.475	3706.581	3953.686	4200.791	4447.897	4695.002
2000	4942.108	5189.213	5436.318	5683.424	5930.529	6177.635	6424.740	6671.845	6918.951	7166.056
3000	7413.161	7660.267	7907.372	8154.478	8401.583	8648.688	8895.794	9142.899	9390.004	9637.110
4000	9884.215	10131.321	10378.426	10625.531	10872.637	11119.742	11366.848	11613.953	11861.058	12108.164

(b)

hectares	0	1	2	3	4	5	6	7	8	9
0	—	2.471	4.942	7.413	9.884	12.355	14.826	17.297	19.768	22.239
10	24.711	27.182	29.653	32.124	34.595	37.066	39.537	42.008	44.479	46.950
20	49.421	51.892	54.363	56.834	59.305	61.776	64.247	66.718	69.190	71.661
30	74.132	76.603	79.074	81.545	84.016	86.487	88.958	91.429	93.900	96.371
40	98.842	101.313	103.784	106.255	108.726	111.197	113.668	116.140	118.611	121.082
50	123.553	126.024	128.495	130.966	133.437	135.908	138.379	140.850	143.321	145.792
60	148.263	150.734	153.205	155.676	158.147	160.618	163.090	165.561	168.032	170.503
70	172.974	175.445	177.916	180.387	182.858	185.329	187.800	190.271	192.742	195.213
80	197.684	200.155	202.626	205.097	207.569	210.040	212.511	214.982	217.453	219.924
90	222.395	224.866	227.337	229.808	232.279	234.750	237.221	239.692	242.163	244.634

(c)

hectares	0.000	0.001	0.002	0.003	0.004	0.005	0.006	0.007	0.008	0.009
0.000	—	0.002	0.005	0.007	0.010	0.012	0.015	0.017	0.020	0.022
0.010	0.025	0.027	0.030	0.032	0.035	0.037	0.040	0.042	0.044	0.047
0.020	0.049	0.052	0.054	0.057	0.059	0.062	0.064	0.067	0.069	0.072
0.030	0.074	0.077	0.079	0.082	0.084	0.086	0.089	0.091	0.094	0.096
0.040	0.099	0.101	0.104	0.106	0.109	0.111	0.114	0.116	0.119	0.121
0.050	0.124	0.126	0.128	0.131	0.133	0.136	0.138	0.141	0.143	0.146
0.060	0.148	0.151	0.153	0.156	0.158	0.161	0.163	0.166	0.168	0.171
0.070	0.173	0.175	0.178	0.180	0.183	0.185	0.188	0.190	0.193	0.195
0.080	0.198	0.200	0.203	0.205	0.208	0.210	0.213	0.215	0.217	0.220
0.090	0.222	0.225	0.227	0.230	0.232	0.235	0.237	0.240	0.242	0.245

hectares	0.000	0.001	0.002	0.003	0.004	0.005	0.006	0.007	0.008	0.009
0.100	0.247	0.250	0.252	0.255	0.257	0.259	0.262	0.264	0.267	0.269
0.110	0.272	0.274	0.277	0.279	0.282	0.284	0.287	0.289	0.292	0.294
0.120	0.297	0.299	0.301	0.304	0.306	0.309	0.311	0.314	0.316	0.319
0.130	0.321	0.324	0.326	0.329	0.331	0.334	0.336	0.339	0.341	0.343
0.140	0.346	0.348	0.351	0.353	0.356	0.358	0.361	0.363	0.366	0.368
0.150	0.371	0.373	0.376	0.378	0.381	0.383	0.385	0.388	0.390	0.393
0.160	0.395	0.398	0.400	0.403	0.405	0.408	0.410	0.413	0.415	0.418
0.170	0.420	0.423	0.425	0.427	0.430	0.432	0.435	0.437	0.440	0.442
0.180	0.445	0.447	0.450	0.452	0.455	0.457	0.460	0.462	0.465	0.467
0.190	0.470	0.472	0.474	0.477	0.479	0.482	0.484	0.487	0.489	0.492
0.200	0.494	0.497	0.499	0.502	0.504	0.507	0.509	0.512	0.514	0.516
0.210	0.519	0.521	0.524	0.526	0.529	0.531	0.534	0.536	0.539	0.541
0.220	0.544	0.546	0.549	0.551	0.554	0.556	0.558	0.561	0.563	0.566
0.230	0.568	0.571	0.573	0.576	0.578	0.581	0.583	0.586	0.588	0.591
0.240	0.593	0.596	0.598	0.600	0.603	0.605	0.608	0.610	0.613	0.615
0.250	0.618	0.620	0.623	0.625	0.628	0.630	0.633	0.635	0.638	0.640
0.260	0.642	0.645	0.647	0.650	0.652	0.655	0.657	0.660	0.662	0.665
0.270	0.667	0.670	0.672	0.675	0.677	0.680	0.682	0.684	0.687	0.689
0.280	0.692	0.694	0.697	0.699	0.702	0.704	0.707	0.709	0.712	0.714
0.290	0.717	0.719	0.722	0.724	0.726	0.729	0.731	0.734	0.736	0.739
0.300	0.741	0.744	0.746	0.749	0.751	0.754	0.756	0.759	0.761	0.764
0.310	0.766	0.768	0.771	0.773	0.776	0.778	0.781	0.783	0.786	0.788
0.320	0.791	0.793	0.796	0.798	0.801	0.803	0.806	0.808	0.811	0.813
0.330	0.815	0.818	0.820	0.823	0.825	0.828	0.830	0.833	0.835	0.838
0.340	0.840	0.843	0.845	0.848	0.850	0.853	0.855	0.857	0.860	0.862
0.350	0.865	0.867	0.870	0.872	0.875	0.877	0.880	0.882	0.885	0.887
0.360	0.890	0.892	0.895	0.897	0.899	0.902	0.904	0.907	0.909	0.912
0.370	0.914	0.917	0.919	0.922	0.924	0.927	0.929	0.932	0.934	0.937
0.380	0.939	0.941	0.944	0.946	0.949	0.951	0.954	0.956	0.959	0.961
0.390	0.964	0.966	0.969	0.971	0.974	0.976	0.979	0.981	0.983	0.986
0.400	0.988	0.991	0.993	0.996	0.998	1.001	1.003	1.006	1.008	1.011
0.410	1.013	1.016	1.018	1.021	1.023	1.025	1.028	1.030	1.033	1.035
0.420	1.038	1.040	1.043	1.045	1.048	1.050	1.053	1.055	1.058	1.060
0.430	1.063	1.065	1.067	1.070	1.072	1.075	1.077	1.080	1.082	1.085
0.440	1.087	1.090	1.092	1.095	1.097	1.100	1.102	1.105	1.107	1.110
0.450	1.112	1.114	1.117	1.119	1.122	1.124	1.127	1.129	1.132	1.134
0.460	1.137	1.139	1.142	1.144	1.147	1.149	1.152	1.154	1.156	1.159
0.470	1.161	1.164	1.166	1.169	1.171	1.174	1.176	1.179	1.181	1.184
0.480	1.186	1.189	1.191	1.194	1.196	1.198	1.201	1.203	1.206	1.208
0.490	1.211	1.213	1.216	1.218	1.221	1.223	1.226	1.228	1.231	1.233
0.500	1.236	1.238	1.240	1.243	1.245	1.248	1.250	1.253	1.255	1.258
0.510	1.260	1.263	1.265	1.268	1.270	1.273	1.275	1.278	1.280	1.282
0.520	1.285	1.287	1.290	1.292	1.295	1.297	1.300	1.302	1.305	1.307
0.530	1.310	1.312	1.315	1.317	1.320	1.322	1.324	1.327	1.329	1.332
0.540	1.334	1.337	1.339	1.342	1.344	1.347	1.349	1.352	1.354	1.357

hectares to acres

hectares	0.000	0.001	0.002	0.003	0.004	0.005	0.006	0.007	0.008	0.009
0.550	1.359	1.362	1.364	1.366	1.369	1.371	1.374	1.376	1.379	1.381
0.560	1.384	1.386	1.389	1.391	1.394	1.396	1.399	1.401	1.404	1.406
0.570	1.409	1.411	1.413	1.416	1.418	1.421	1.423	1.426	1.428	1.431
0.580	1.433	1.436	1.438	1.441	1.443	1.446	1.448	1.451	1.453	1.455
0.590	1.458	1.460	1.463	1.465	1.468	1.470	1.473	1.475	1.478	1.480
0.600	1.483	1.485	1.488	1.490	1.493	1.495	1.497	1.500	1.502	1.505
0.610	1.507	1.510	1.512	1.515	1.517	1.520	1.522	1.525	1.527	1.530
0.620	1.532	1.535	1.537	1.539	1.542	1.544	1.547	1.549	1.552	1.554
0.630	1.557	1.559	1.562	1.564	1.567	1.569	1.572	1.574	1.577	1.579
0.640	1.581	1.584	1.586	1.589	1.591	1.594	1.596	1.599	1.601	1.604
0.650	1.606	1.609	1.611	1.614	1.616	1.619	1.621	1.623	1.626	1.628
0.660	1.631	1.633	1.636	1.638	1.641	1.643	1.646	1.648	1.651	1.653
0.670	1.656	1.658	1.661	1.663	1.665	1.668	1.670	1.673	1.675	1.678
0.680	1.680	1.683	1.685	1.688	1.690	1.693	1.695	1.698	1.700	1.703
0.690	1.705	1.707	1.710	1.712	1.715	1.717	1.720	1.722	1.725	1.727
0.700	1.730	1.732	1.735	1.737	1.740	1.742	1.745	1.747	1.750	1.752
0.710	1.754	1.757	1.759	1.762	1.764	1.767	1.769	1.772	1.774	1.777
0.720	1.779	1.782	1.784	1.787	1.789	1.792	1.794	1.796	1.799	1.801
0.730	1.804	1.806	1.809	1.811	1.814	1.816	1.819	1.821	1.824	1.826
0.740	1.829	1.831	1.834	1.836	1.838	1.841	1.843	1.846	1.848	1.851
0.750	1.853	1.856	1.858	1.861	1.863	1.866	1.868	1.871	1.873	1.876
0.760	1.878	1.880	1.883	1.885	1.888	1.890	1.893	1.895	1.898	1.900
0.770	1.903	1.905	1.908	1.910	1.913	1.915	1.918	1.920	1.922	1.925
0.780	1.927	1.930	1.932	1.935	1.937	1.940	1.942	1.945	1.947	1.950
0.790	1.952	1.955	1.957	1.960	1.962	1.964	1.967	1.969	1.972	1.974
0.800	1.977	1.979	1.982	1.984	1.987	1.989	1.992	1.994	1.997	1.999
0.810	2.002	2.004	2.006	2.009	2.011	2.014	2.016	2.019	2.021	2.024
0.820	2.026	2.029	2.031	2.034	2.036	2.039	2.041	2.044	2.046	2.049
0.830	2.051	2.053	2.056	2.058	2.061	2.063	2.066	2.068	2.071	2.073
0.840	2.076	2.078	2.081	2.083	2.086	2.088	2.091	2.093	2.095	2.098
0.850	2.100	2.103	2.105	2.108	2.110	2.113	2.115	2.118	2.120	2.123
0.860	2.125	2.128	2.130	2.133	2.135	2.137	2.140	2.142	2.145	2.147
0.870	2.150	2.152	2.155	2.157	2.160	2.162	2.165	2.167	2.170	2.172
0.880	2.175	2.177	2.179	2.182	2.184	2.187	2.189	2.192	2.194	2.197
0.890	2.199	2.202	2.204	2.207	2.209	2.212	2.214	2.217	2.219	2.221
0.900	2.224	2.226	2.229	2.231	2.234	2.236	2.239	2.241	2.244	2.246
0.910	2.249	2.251	2.254	2.256	2.259	2.261	2.263	2.266	2.268	2.271
0.920	2.273	2.276	2.278	2.281	2.283	2.286	2.288	2.291	2.293	2.296
0.930	2.298	2.301	2.303	2.305	2.308	2.310	2.313	2.315	2.318	2.320
0.940	2.323	2.325	2.328	2.330	2.333	2.335	2.338	2.340	2.343	2.345
0.950	2.348	2.350	2.352	2.355	2.357	2.360	2.362	2.365	2.367	2.370
0.960	2.372	2.375	2.377	2.380	2.382	2.385	2.387	2.390	2.392	2.394
0.970	2.397	2.399	2.402	2.404	2.407	2.409	2.412	2.414	2.417	2.419
0.980	2.422	2.424	2.427	2.429	2.432	2.434	2.436	2.439	2.441	2.444
0.990	2.446	2.449	2.451	2.454	2.456	2.459	2.461	2.464	2.466	2.469

Example: 3456.7 ha = 8401.583 + 138.379 + 1.730 acres = 8541.692 acres

Table 23 SQUARE MILES TO SQUARE KILOMETRES

(a)
1 mile² = 2.589 988 11 km²

square miles	0	100	200	300	400	500	600	700	800	900
0	—	258.999	517.998	776.996	1035.995	1294.994	1553.993	1812.992	2071.990	2330.989
1000	2589.988	2848.987	3107.986	3366.985	3625.983	3884.982	4143.981	4402.980	4661.979	4920.977
2000	5179.976	5438.975	5697.974	5956.973	6215.971	6474.970	6733.969	6992.968	7251.967	7510.966
3000	7769.964	8028.963	8287.962	8546.961	8805.960	9064.958	9323.957	9582.956	9841.955	10100.954

(b)

square miles	0	1	2	3	4	5	6	7	8	9
0	—	2.590	5.180	7.770	10.360	12.950	15.540	18.130	20.720	23.310
10	25.900	28.490	31.080	33.670	36.260	38.850	41.440	44.030	46.620	49.210
20	51.800	54.390	56.980	59.570	62.160	64.750	67.340	69.930	72.520	75.110
30	77.700	80.290	82.880	85.470	88.060	90.650	93.240	95.830	98.420	101.010
40	103.600	106.190	108.780	111.369	113.959	116.549	119.139	121.729	124.319	126.909
50	129.499	132.089	134.679	137.269	139.859	142.449	145.039	147.629	150.219	152.809
60	155.399	157.989	160.579	163.169	165.759	168.349	170.939	173.529	176.119	178.709
70	181.299	183.889	186.479	189.069	191.659	194.249	196.839	199.429	202.019	204.609
80	207.199	209.789	212.379	214.969	217.559	220.149	222.739	225.329	227.919	230.509
90	233.099	235.689	238.279	240.869	243.459	246.049	248.639	251.229	253.819	256.409

Example: 1903 mile² = 4920.977 + 7.770 km² = 4928.747 km²

Table 24 SQUARE KILOMETRES TO SQUARE MILES

(a)
1 km² = 0.386 102 159 mile²

square kilometres	0	100	200	300	400	500	600	700	800	900
0	—	38.6102	77.2204	115.8306	154.4409	193.0511	231.6613	270.2715	308.8817	347.4919
1000	386.1022	424.7124	463.3226	501.9328	540.5430	579.1532	617.7635	656.3737	694.9839	733.5941
2000	772.2043	810.8145	849.4247	888.0350	926.6452	965.2554	1003.8656	1042.4758	1081.0860	1119.6963
3000	1158.3065	1196.9167	1235.5269	1274.1371	1312.7473	1351.3576	1389.9678	1428.5780	1467.1882	1505.7984
4000	1544.4086	1583.0189	1621.6291	1660.2393	1698.8495	1737.4597	1776.0699	1814.6801	1853.2904	1891.9006

(b)

square kilometres	0	1	2	3	4	5	6	7	8	9
0	—	0.3861	0.7722	1.1583	1.5444	1.9305	2.3166	2.7027	3.0888	3.4749
10	3.8610	4.2471	4.6332	5.0193	5.4054	5.7915	6.1776	6.5637	6.9498	7.3359
20	7.7220	8.1081	8.4942	8.8803	9.2665	9.6526	10.0387	10.4248	10.8109	11.1970
30	11.5831	11.9692	12.3553	12.7414	13.1275	13.5136	13.8997	14.2858	14.6719	15.0580
40	15.4441	15.8302	16.2163	16.6024	16.9885	17.3746	17.7607	18.1468	18.5329	18.9190
50	19.3051	19.6912	20.0773	20.4634	20.8495	21.2356	21.6217	22.0078	22.3939	22.7800
60	23.1661	23.5522	23.9383	24.3244	24.7105	25.0966	25.4827	25.8688	26.2549	26.6410
70	27.0272	27.4133	27.7994	28.1855	28.5716	28.9577	29.3438	29.7299	30.1160	30.5021
80	30.8882	31.2743	31.6604	32.0465	32.4326	32.8187	33.2048	33.5909	33.9770	34.3631
90	34.7492	35.1353	35.5214	35.9075	36.2936	36.6797	37.0658	37.4519	37.8380	38.2241

Example: 2761 km² = 1042.4758 + 23.5522 mile² = 1066.0280 mile²
Population Densities — To convert persons per square mile to persons per square kilometre (and vice versa), see Example 4 on page 7

Table 25 ACRES TO SQUARE MILES

1 acre = 0.001 562 5 square mile (exactly)
All values in Table 25(a) are exact

(a)

acres	0	100000	200000	300000	400000	500000	600000	700000	800000	900000
0	0	156.250	312.500	468.750	625.000	781.250	937.500	1093.75	1250.00	1406.25
1000000	1562.50	1718.75	1875.00	2031.25	2187.50	2343.75	2500.00	2656.25	2812.50	2968.75
2000000	3125.00	3281.25	3437.50	3593.75	3750.00	3906.25	4062.50	4218.75	4375.00	4531.25

(b)

acres	0	1000	2000	3000	4000	5000	6000	7000	8000	9000
0	—	1.56	3.13	4.69	6.25	7.81	9.38	10.94	12.50	14.06
10000	15.63	17.19	18.75	20.31	21.88	23.44	25.00	26.56	28.13	29.69
20000	31.25	32.81	34.38	35.94	37.50	39.06	40.63	42.19	43.75	45.31
30000	46.88	48.44	50.00	51.56	53.13	54.69	56.25	57.81	59.38	60.94
40000	62.50	64.06	65.63	67.19	68.75	70.31	71.88	73.44	75.00	76.56
50000	78.13	79.69	81.25	82.81	84.38	85.94	87.50	89.06	90.63	92.19
60000	93.75	95.31	96.88	98.44	100.00	101.56	103.13	104.69	106.25	107.81
70000	109.38	110.94	112.50	114.06	115.63	117.19	118.75	120.31	121.88	123.44
80000	125.00	126.56	128.13	129.69	131.25	132.81	134.38	135.94	137.50	139.06
90000	140.63	142.19	143.75	145.31	146.88	148.44	150.00	151.56	153.13	154.69

Example: 1111000 acres = 1718.75 + 17.19 miles^2 = 1735.94 miles^2

Table 26 SQUARE MILES TO ACRES

1 square mile = 640 acres (exactly)
All values in Table 26(a) and (b) are exact

(a)

square miles	0	100	200	300	400	500	600	700	800	900
0	—	64000	128000	192000	256000	320000	384000	448000	512000	576000
1000	640000	704000	768000	832000	896000	960000	1024000	1088000	1152000	1216000
2000	1280000	1344000	1408000	1472000	1536000	1600000	1664000	1728000	1792000	1856000

(b)

square miles	0	1	2	3	4	5	6	7	8	9
0	—	640	1280	1920	2560	3200	3840	4480	5120	5760
10	6400	7040	7680	8320	8960	9600	10240	10880	11520	12160
20	12800	13440	14080	14720	15360	16000	16640	17280	17920	18560
30	19200	19840	20480	21120	21760	22400	23040	23680	24320	24960
40	25600	26240	26880	27520	28160	28800	29440	30080	30720	31360
50	32000	32640	33280	33920	34560	35200	35840	36480	37120	37760
60	38400	39040	39680	40320	40960	41600	42240	42880	43520	44160
70	44800	45440	46080	46720	47360	48000	48640	49280	49920	50560
80	51200	51840	52480	53120	53760	54400	55040	55680	56320	56960
90	57600	58240	58880	59520	60160	60800	61440	62080	62720	63360

Example: 2622 miles^2 = 1664000 + 14080 acres = 1678080 acres

Table 27 CUBIC INCHES TO CUBIC CENTIMETRES

[in³ to cm³]   1 in³ = 16.387 064 cm³ (exactly)

(a)

cubic inches	0	100	200	300	400	500	600	700	800	900
0	—	1638.71	3277.41	4916.12	6554.83	8193.53	9832.24	11470.94	13109.65	14748.36

(b)

cubic inches	0	1	2	3	4	5	6	7	8	9
0	—	16.39	32.77	49.16	65.55	81.94	98.32	114.71	131.10	147.48
10	163.87	180.26	196.64	213.03	229.42	245.81	262.19	278.58	294.97	311.35
20	327.74	344.13	360.52	376.90	393.29	409.68	426.06	442.45	458.84	475.22
30	491.61	508.00	524.39	540.77	557.16	573.55	589.93	606.32	622.71	639.10
40	655.48	671.87	688.26	704.64	721.03	737.42	753.80	770.19	786.58	802.97
50	819.35	835.74	852.13	868.51	884.90	901.29	917.68	934.06	950.45	966.84
60	983.22	999.61	1016.00	1032.39	1048.77	1065.16	1081.55	1097.93	1114.32	1130.71
70	1147.09	1163.48	1179.87	1196.26	1212.64	1229.03	1245.42	1261.80	1278.19	1294.58
80	1310.97	1327.35	1343.74	1360.13	1376.51	1392.90	1409.29	1425.67	1442.06	1458.45
90	1474.84	1491.22	1507.61	1524.00	1540.38	1556.77	1573.16	1589.55	1605.93	1622.32

Example: 871 in³ = 13109.65 + 1163.48 cm³ = 14273.13 cm³

Note: In order to convert:
  (i) cubic inches to cubic millimetres, shift the decimal point in Table 27 three places to the right;
  (ii) cubic inches to cubic decimetres, shift the decimal point in Table 27 three places to the left.

Table 28 CUBIC CENTIMETRES TO CUBIC INCHES

[cm³ to in³]   1 cm³ = 0.061 023 744 1 in³

(a)

cubic centimetres	0	100	200	300	400	500	600	700	800	900
0	—	6.1024	12.2047	18.3071	24.4095	30.5119	36.6142	42.7166	48.8190	54.9214

(b)

cubic centimetres	0	1	2	3	4	5	6	7	8	9
0	—	0.0610	0.1220	0.1831	0.2441	0.3051	0.3661	0.4272	0.4882	0.5492
10	0.6102	0.6713	0.7323	0.7933	0.8543	0.9154	0.9764	1.0374	1.0984	1.1595
20	1.2205	1.2815	1.3425	1.4035	1.4646	1.5256	1.5866	1.6476	1.7087	1.7697
30	1.8307	1.8917	1.9528	2.0138	2.0748	2.1358	2.1969	2.2579	2.3189	2.3799
40	2.4409	2.5020	2.5630	2.6240	2.6850	2.7461	2.8071	2.8681	2.9291	2.9902
50	3.0512	3.1122	3.1732	3.2343	3.2953	3.3563	3.4173	3.4784	3.5394	3.6004
60	3.6614	3.7224	3.7835	3.8445	3.9055	3.9665	4.0276	4.0886	4.1496	4.2106
70	4.2717	4.3327	4.3937	4.4547	4.5158	4.5768	4.6378	4.6988	4.7599	4.8209
80	4.8819	4.9429	5.0039	5.0650	5.1260	5.1870	5.2480	5.3091	5.3701	5.4311
90	5.4921	5.5532	5.6142	5.6752	5.7362	5.7973	5.8583	5.9193	5.9803	6.0414

Example: 723 cm³ = 42.7166 + 1.4035 in³ = 44.1201 in³

Table 29 CUBIC FEET TO CUBIC METRES

[ft³ to m³]   1 ft³ = 0.028 316 846 592 m³ (exactly)

(a)

cubic feet	0	100	200	300	400	500	600	700	800	900
0	—	2.83168	5.66337	8.49505	11.32674	14.15842	16.99011	19.82179	22.65348	25.48516
1000	28.31685	31.14853	33.98022	36.81190	39.64359	42.47527	45.30695	48.13864	50.97032	53.80201
2000	56.63369	59.46538	62.29706	65.12875	67.96043	70.79212	73.62380	76.45549	79.28717	82.11886

(b)

cubic feet	0	1	2	3	4	5	6	7	8	9
0	—	0.02832	0.05663	0.08495	0.11327	0.14158	0.16990	0.19822	0.22653	0.25485
10	0.28317	0.31149	0.33980	0.36812	0.39644	0.42475	0.45307	0.48139	0.50970	0.53802
20	0.56634	0.59465	0.62297	0.65129	0.67960	0.70792	0.73624	0.76455	0.79287	0.82119
30	0.84951	0.87782	0.90614	0.93446	0.96277	0.99109	1.01941	1.04772	1.07604	1.10436
40	1.13267	1.16099	1.18931	1.21762	1.24594	1.27426	1.30257	1.33089	1.35921	1.38753
50	1.41584	1.44416	1.47248	1.50079	1.52911	1.55743	1.58574	1.61406	1.64238	1.67069
60	1.69901	1.72733	1.75564	1.78396	1.81228	1.84060	1.86891	1.89723	1.92555	1.95386
70	1.98218	2.01050	2.03881	2.06713	2.09545	2.12376	2.15208	2.18040	2.20871	2.23703
80	2.26535	2.29366	2.32198	2.35030	2.37862	2.40693	2.43525	2.46357	2.49188	2.52020
90	2.54852	2.57683	2.60515	2.63347	2.66178	2.69010	2.71842	2.74673	2.77505	2.80337

Example: 2671 ft³ = 73.62380 + 2.01050 m³ = 75.63430 m³
Note: In order to convert cubic feet to cubic decimetres, shift the decimal point in Table 29 three places to the right.

Table 30 CUBIC METRES TO CUBIC FEET

[m³ to ft³]   1 m³ = 35.314 666 7 ft³

(a)

cubic metres	0	100	200	300	400	500	600	700	800	900
0	—	3531.47	7062.93	10594.40	14125.87	17657.33	21188.80	24720.27	28251.73	31783.20
1000	35314.67	38846.13	42377.60	45909.07	49440.53	52972.00	56503.47	60034.93	63566.40	67097.87
2000	70629.33	74160.80	77692.27	81223.73	84755.20	88286.67	91818.13	95349.60	98881.07	102412.53

(b)

cubic metres	0	1	2	3	4	5	6	7	8	9
0	—	35.31	70.63	105.94	141.26	176.57	211.89	247.20	282.52	317.83
10	353.15	388.46	423.78	459.09	494.41	529.72	565.03	600.35	635.66	670.98
20	706.29	741.61	776.92	812.24	847.55	882.87	918.18	953.50	988.81	1024.13
30	1059.44	1094.75	1130.07	1165.38	1200.70	1236.01	1271.33	1306.64	1341.96	1377.27
40	1412.59	1447.90	1483.22	1518.53	1553.85	1589.16	1624.47	1659.79	1695.10	1730.42
50	1765.73	1801.05	1836.36	1871.68	1906.99	1942.31	1977.62	2012.94	2048.25	2083.57
60	2118.88	2154.19	2189.51	2224.82	2260.14	2295.45	2330.77	2366.08	2401.40	2436.71
70	2472.03	2507.34	2542.66	2577.97	2613.29	2648.60	2683.91	2719.23	2754.54	2789.86
80	2825.17	2860.49	2895.80	2931.12	2966.43	3001.75	3037.06	3072.38	3107.69	3143.01
90	3178.32	3213.63	3248.95	3284.26	3319.58	3354.89	3390.21	3425.52	3460.84	3496.15

Example: 633 m³ = 21188.80 + 1165.38 ft³ = 22354.18 ft³
Note: the above tables can also be used to determine *Volume Rate of Flow*, e.g. cubic feet per second (cusec) to cubic metres per second, and cubic metres per second to cubic feet per second (cusec).

Table 31 CUBIC YARDS TO CUBIC METRES

[yd³ to m³]   1 yd³ = 0.764 554 857 984 m³ (exactly)

(a)

cubic yards	0	100	200	300	400	500	600	700	800	900
0	—	76.4555	152.9110	229.3665	305.8219	382.2774	458.7329	535.1884	611.6439	688.0994
1000	764.5549	841.0103	917.4658	993.9213	1070.3768	1146.8323	1223.2878	1299.7433	1376.1987	1452.6542
2000	1529.1097	1605.5652	1682.0207	1758.4762	1834.9317	1911.3871	1987.8426	2064.2981	2140.7536	2217.2091

(b)

cubic yards	0	1	2	3	4	5	6	7	8	9
0	—	0.7646	1.5291	2.2937	3.0582	3.8228	4.5873	5.3519	6.1164	6.8810
10	7.6455	8.4101	9.1747	9.9392	10.7038	11.4683	12.2329	12.9974	13.7620	14.5265
20	15.2911	16.0557	16.8202	17.5848	18.3493	19.1139	19.8784	20.6430	21.4075	22.1721
30	22.9366	23.7012	24.4658	25.2303	25.9949	26.7594	27.5240	28.2885	29.0531	29.8176
40	30.5822	31.3467	32.1113	32.8759	33.6404	34.4050	35.1695	35.9341	36.6986	37.4632
50	38.2277	38.9923	39.7569	40.5214	41.2860	42.0505	42.8151	43.5796	44.3442	45.1087
60	45.8733	46.6378	47.4024	48.1670	48.9315	49.6961	50.4606	51.2252	51.9897	52.7543
70	53.5188	54.2834	55.0479	55.8125	56.5771	57.3416	58.1062	58.8707	59.6353	60.3998
80	61.1644	61.9289	62.6935	63.4581	64.2226	64.9872	65.7517	66.5163	67.2808	68.0454
90	68.8099	69.5745	70.3390	71.1036	71.8682	72.6327	73.3973	74.1618	74.9264	75.6909

Example: 1044 yd³ = 764.5549 + 33.6404 m³ = 798.1953 m³

Table 32 CUBIC METRES TO CUBIC YARDS

[m³ to yd³]   1 m³ = 1.307 950 62 yd³

(a)

cubic metres	0	100	200	300	400	500	600	700	800	900
0	—	130.7951	261.5901	392.3852	523.1802	653.9753	784.7704	915.5654	1046.3605	1177.1556
1000	1307.9506	1438.7457	1569.5407	1700.3358	1831.1309	1961.9259	2092.7210	2223.5161	2354.3111	2485.1062
2000	2615.9012	2746.6963	2877.4914	3008.2864	3139.0815	3269.8766	3400.6716	3531.4667	3662.2617	3793.0568

(b)

cubic metres	0	1	2	3	4	5	6	7	8	9
0	—	1.3080	2.6159	3.9239	5.2318	6.5398	7.8477	9.1557	10.4636	11.7716
10	13.0795	14.3875	15.6954	17.0034	18.3113	19.6193	20.9272	22.2352	23.5431	24.8511
20	26.1590	27.4670	28.7749	30.0829	31.3908	32.6988	34.0067	35.3147	36.6226	37.9306
30	39.2385	40.5465	41.8544	43.1624	44.4703	45.7783	47.0862	48.3942	49.7021	51.0101
40	52.3180	53.6260	54.9339	56.2419	57.5498	58.8578	60.1657	61.4737	62.7816	64.0896
50	65.3975	66.7055	68.0134	69.3214	70.6293	71.9373	73.2452	74.5532	75.8611	77.1691
60	78.4770	79.7850	81.0929	82.4009	83.7088	85.0168	86.3247	87.6327	88.9406	90.2486
70	91.5565	92.8645	94.1724	95.4804	96.7883	98.0963	99.4042	100.7122	102.0201	103.3281
80	104.6360	105.9440	107.2520	108.5599	109.8679	111.1758	112.4838	113.7917	115.0997	116.4076
90	117.7156	119.0235	120.3315	121.6394	122.9474	124.2553	125.5633	126.8712	128.1792	129.4871

Example: 2702 m³ = 3531.4667 + 2.6159 yd³ = 3534.0826 yd³

Table 33 UK GALLONS TO LITRES
[UKgal to l]   1 UKgal = 4.546 091 878 l

(a)

UK gallons	0	100	200	300	400	500	600	700	800	900
0	—	454.609	909.218	1363.828	1818.437	2273.046	2727.655	3182.264	3636.874	4091.483
1000	4546.092	5000.701	5455.310	5909.919	6364.529	6819.138	7273.747	7728.356	8182.965	8637.575
2000	9092.184	9546.793	10001.402	10456.011	10910.621	11365.230	11819.839	12274.448	12729.057	13183.666

(b)

UK gallons	0	1	2	3	4	5	6	7	8	9
0	—	4.546	9.092	13.638	18.184	22.730	27.277	31.823	36.369	40.915
10	45.461	50.007	54.553	59.099	63.645	68.191	72.737	77.284	81.830	86.376
20	90.922	95.468	100.014	104.560	109.106	113.652	118.198	122.744	127.291	131.837
30	136.383	140.929	145.475	150.021	154.567	159.113	163.659	168.205	172.751	177.298
40	181.844	186.390	190.936	195.482	200.028	204.574	209.120	213.666	218.212	222.759
50	227.305	231.851	236.397	240.943	245.489	250.035	254.581	259.127	263.673	268.219
60	272.766	277.312	281.858	286.404	290.950	295.496	300.042	304.588	309.134	313.680
70	318.226	322.773	327.319	331.865	336.411	340.957	345.503	350.049	354.595	359.141
80	363.687	368.233	372.780	377.326	381.872	386.418	390.964	395.510	400.056	404.602
90	409.148	413.694	418.240	422.787	427.333	431.879	436.425	440.971	445.517	450.063

Example: 2484 UKgal = 10910.621 + 381.872 l = 11292.493 l

Note on the litre and the UK gallon
The litre used in the tables in this book equals one cubic decimetre as redefined by the 12th International Conference on Weights and Measures, New Delhi, 1964, (1 UK gallon = 4.546 091 878 litres) which it is anticipated will soon be legally recognised in the United Kingdom. At the date of publication (1973), however, the litre legally recognised in the United Kingdom is the one authorised by the Weights and Measures Act, 1963, namely 1 UK gallon = 4.545 964 591 litres.

It should be noted that these two conversion factors are identical to four significant figures. Tables 33 and 34 can therefore be used to three significant figures where l UK gallon = 4.545 964 591 litres.
Note: ½ UKgal = 4 pints = 2.273 l; ¼ UKgal = 2 pints = 1.137 l; ⅛ UKgal = 1 pint = 0.568 l

Table 34 LITRES TO UK GALLONS
[l to UKgal]   1 l = 0.219 969 157 UKgal

(a)

litres	0	100	200	300	400	500	600	700	800	900
0	—	21.9969	43.9938	65.9907	87.9877	109.9846	131.9815	153.9784	175.9753	197.9722
1000	219.9692	241.9661	263.9630	285.9599	307.9568	329.9537	351.9507	373.9476	395.9445	417.9414
2000	439.9383	461.9352	483.9321	505.9291	527.9260	549.9229	571.9198	593.9167	615.9136	637.9106
3000	659.9075	681.9044	703.9013	725.8982	747.8951	769.8920	791.8890	813.8859	835.8828	857.8797
4000	879.8766	901.8735	923.8705	945.8674	967.8643	989.8612	1011.8581	1033.8550	1055.8520	1077.8489

(b)

litres	0	1	2	3	4	5	6	7	8	9
0	—	0.2200	0.4399	0.6599	0.8799	1.0998	1.3198	1.5398	1.7598	1.9797
10	2.1997	2.4197	2.6396	2.8596	3.0796	3.2995	3.5195	3.7395	3.9594	4.1794
20	4.3994	4.6194	4.8393	5.0593	5.2793	5.4992	5.7192	5.9392	6.1591	6.3791
30	6.5991	6.8190	7.0390	7.2590	7.4790	7.6989	7.9189	8.1389	8.3588	8.5788
40	8.7988	9.0187	9.2387	9.4587	9.6786	9.8986	10.1186	10.3386	10.5585	10.7785
50	10.9985	11.2184	11.4384	11.6584	11.8783	12.0983	12.3183	12.5382	12.7582	12.9782
60	13.1981	13.4181	13.6381	13.8581	14.0780	14.2980	14.5180	14.7379	14.9579	15.1779
70	15.3978	15.6178	15.8378	16.0577	16.2777	16.4977	16.7177	16.9376	17.1576	17.3776
80	17.5975	17.8175	18.0375	18.2574	18.4774	18.6974	18.9173	19.1373	19.3573	19.5773
90	19.7972	20.0172	20.2372	20.4571	20.6771	20.8971	21.1170	21.3370	21.5570	21.7769

Example: 3456 l = 747.8951 + 12.3183 UKgal = 760.2134 UKgal
Note: 1 l = 1.7598 UK pints = 1¾ pints (approx.)

Table 35 US GALLONS TO LITRES

[USgal to l]   1 USgal = 3.785 411 784 l

(a)

US gallons	0	100	200	300	400	500	600	700	800	900
0	—	378.541	757.082	1135.624	1514.165	1892.706	2271.247	2649.788	3028.329	3406.871
1000	3785.412	4163.953	4542.494	4921.035	5299.576	5678.118	6056.659	6435.200	6813.741	7192.282
2000	7570.824	7949.365	8327.906	8706.447	9084.988	9463.529	9842.071	10220.612	10599.153	10977.694

(b)

US gallons	0	1	2	3	4	5	6	7	8	9
0	—	3.785	7.571	11.356	15.142	18.927	22.712	26.498	30.283	34.069
10	37.854	41.640	45.425	49.210	52.996	56.781	60.567	64.352	68.137	71.923
20	75.708	79.494	83.279	87.064	90.850	94.635	98.421	102.206	105.992	109.777
30	113.562	117.348	121.133	124.919	128.704	132.489	136.275	140.060	143.846	147.631
40	151.416	155.202	158.987	162.773	166.558	170.344	174.129	177.914	181.700	185.485
50	189.271	193.056	196.841	200.627	204.412	208.198	211.983	215.768	219.554	223.339
60	227.125	230.910	234.696	238.481	242.266	246.052	249.837	253.623	257.408	261.193
70	264.979	268.764	272.550	276.335	280.120	283.906	287.691	291.477	295.262	299.048
80	302.833	306.618	310.404	314.189	317.975	321.760	325.545	329.331	333.116	336.902
90	340.687	344.472	348.258	352.043	355.829	359.614	363.400	367.185	370.970	374.756

Example: 2863 USgal = 10599.153 + 238.481 l = 10837.634 l

Table 36 LITRES TO US GALLONS

[l to USgal]   1 l = 0.264 172 052 USgal

(a)

litres	0	100	200	300	400	500	600	700	800	900
0	—	26.4172	52.8344	79.2516	105.6688	132.0860	158.5032	184.9204	211.3376	237.7548
1000	264.1721	290.5893	317.0065	343.4237	369.8409	396.2581	422.6753	449.0925	475.5097	501.9269
2000	528.3441	554.7613	581.1785	607.5957	634.0129	660.4301	686.8473	713.2645	739.6817	766.0990
3000	792.5162	818.9334	845.3506	871.7678	898.1850	924.6022	951.0194	977.4366	1003.8538	1030.2710
4000	1056.6882	1083.1054	1109.5226	1135.9398	1162.3570	1188.7742	1215.1914	1241.6086	1268.0258	1294.4431

(b)

litres	0	1	2	3	4	5	6	7	8	9
0	—	0.2642	0.5283	0.7925	1.0567	1.3209	1.5850	1.8492	2.1134	2.3775
10	2.6417	2.9059	3.1701	3.4342	3.6984	3.9626	4.2268	4.4909	4.7551	5.0193
20	5.2834	5.5476	5.8118	6.0760	6.3401	6.6043	6.8685	7.1326	7.3968	7.6610
30	7.9252	8.1893	8.4535	8.7177	8.9818	9.2460	9.5102	9.7744	10.0385	10.3027
40	10.5669	10.8311	11.0952	11.3594	11.6236	11.8877	12.1519	12.4161	12.6803	12.9444
50	13.2086	13.4728	13.7369	14.0011	14.2653	14.5295	14.7936	15.0578	15.3220	15.5862
60	15.8503	16.1145	16.3787	16.6428	16.9070	17.1712	17.4354	17.6995	17.9637	18.2279
70	18.4920	18.7562	19.0204	19.2846	19.5487	19.8129	20.0771	20.3412	20.6054	20.8696
80	21.1338	21.3979	21.6621	21.9263	22.1905	22.4546	22.7188	22.9830	23.2471	23.5113
90	23.7755	24.0397	24.3038	24.5680	24.8322	25.0963	25.3605	25.6247	25.8889	26.1530

Example: 1972 l = 501.9269 + 19.0204 USgal = 520.9473 USgal

Table 37 UK BUSHELS TO HECTOLITRES
1 UK bushel = 0.363 687 350 hl

(a)

UK bushels	0	100	200	300	400	500	600	700	800	900
0	—	36.3687	72.7375	109.1062	145.4749	181.8437	218.2124	254.5811	290.9499	327.31
1000	363.6874	400.0561	436.4248	472.7936	509.1623	545.5310	581.8998	618.2685	654.6372	691.00
2000	727.3747	763.7434	800.1122	836.4809	872.8496	909.2184	945.5871	981.9558	1018.3246	1054.69
3000	1091.0621	1127.4308	1163.7995	1200.1683	1236.5370	1272.9057	1309.2745	1345.6432	1382.0119	1418.38
4000	1454.7494	1491.1181	1527.4869	1563.8556	1600.2243	1636.5931	1672.9618	1709.3305	1745.6993	1782.0

(b)

UK bushels	0	1	2	3	4	5	6	7	8	9
0	—	0.3637	0.7274	1.0911	1.4547	1.8184	2.1821	2.5458	2.9095	3.2?
10	3.6369	4.0006	4.3642	4.7279	5.0916	5.4553	5.8190	6.1827	6.5464	6.9?
20	7.2737	7.6374	8.0011	8.3648	8.7285	9.0922	9.4559	9.8196	10.1832	10.5?
30	10.9106	11.2743	11.6380	12.0017	12.3654	12.7291	13.0927	13.4564	13.8201	14.1?
40	14.5475	14.9112	15.2749	15.6386	16.0022	16.3659	16.7296	17.0933	17.4570	17.82
50	18.1844	18.5481	18.9117	19.2754	19.6391	20.0028	20.3665	20.7302	21.0939	21.4?
60	21.8212	22.1849	22.5486	22.9123	23.2760	23.6397	24.0034	24.3671	24.7307	25.0?
70	25.4581	25.8218	26.1855	26.5492	26.9129	27.2766	27.6402	28.0039	28.3676	28.7?
80	29.0950	29.4587	29.8224	30.1861	30.5497	30.9134	31.2771	31.6408	32.0045	32.3?
90	32.7319	33.0955	33.4592	33.8229	34.1866	34.5503	34.9140	35.2777	35.6414	36.00

Example: 3457 UK bushels = 1236.5370 + 20.7302 hl = 1257.2672 hl

Table 38 HECTOLITRES TO UK BUSHELS
1 hl = 2.749 614 47 UK bushels

(a)

hecto-litres	0	100	200	300	400	500	600	700	800	900
0	—	274.961	549.923	824.884	1099.846	1374.807	1649.769	1924.730	2199.692	2474.?
1000	2749.614	3024.576	3299.537	3574.499	3849.460	4124.422	4399.383	4674.345	4949.306	5224.?
2000	5499.229	5774.190	6049.152	6324.113	6599.075	6874.036	7148.998	7423.959	7698.921	7973.?

(b)

hecto-litres	0	1	2	3	4	5	6	7	8	9
0	—	2.750	5.499	8.249	10.998	13.748	16.498	19.247	21.997	24.?
10	27.496	30.246	32.995	35.745	38.495	41.244	43.994	46.743	49.493	52.?
20	54.992	57.742	60.492	63.241	65.991	68.740	71.490	74.240	76.989	79.?
30	82.488	85.238	87.988	90.737	93.487	96.237	98.986	101.736	104.485	107.?
40	109.985	112.734	115.484	118.233	120.983	123.733	126.482	129.232	131.981	134.?
50	137.481	140.230	142.980	145.730	148.479	151.229	153.978	156.728	159.478	162.?
60	164.977	167.726	170.476	173.226	175.975	178.725	181.475	184.224	186.974	189.?
70	192.473	195.223	197.972	200.722	203.471	206.221	208.971	211.720	214.470	217.?
80	219.969	222.719	225.468	228.218	230.968	233.717	236.467	239.216	241.966	244.?
90	247.465	250.215	252.965	255.714	258.464	261.213	263.963	266.713	269.462	272.?

Example: 2673 hl = 7148.998 + 200.722 UK bushels = 7349.720 UK bushels

Table 39 US BUSHELS TO HECTOLITRES

1 US bushel = 0.352 390 702 hl

US bushels	0	100	200	300	400	500	600	700	800	900
0	—	35.2391	70.4781	105.7172	140.9563	176.1954	211.4344	246.6735	281.9126	317.1516
1000	352.3907	387.6298	422.8688	458.1079	493.3470	528.5861	563.8251	599.0642	634.3033	669.5423
2000	704.7814	740.0205	775.2595	810.4986	845.7377	880.9768	916.2158	951.4549	986.6940	1021.9330
3000	1057.1721	1092.4112	1127.6502	1162.8893	1198.1284	1233.3675	1268.6065	1303.8456	1339.0847	1374.3237
4000	1409.5628	1444.8019	1480.0409	1515.2800	1550.5191	1585.7582	1620.9972	1656.2363	1691.4754	1726.7144

US bushels	0	1	2	3	4	5	6	7	8	9
0	—	0.3524	0.7048	1.0572	1.4096	1.7620	2.1143	2.4667	2.8191	3.1715
10	3.5239	3.8763	4.2287	4.5811	4.9335	5.2859	5.6383	5.9906	6.3430	6.6954
20	7.0478	7.4002	7.7526	8.1050	8.4574	8.8098	9.1622	9.5145	9.8669	10.2193
30	10.5717	10.9241	11.2765	11.6289	11.9813	12.3337	12.6861	13.0385	13.3908	13.7432
40	14.0956	14.4480	14.8004	15.1528	15.5052	15.8576	16.2100	16.5624	16.9148	17.2671
50	17.6195	17.9719	18.3243	18.6767	19.0291	19.3815	19.7339	20.0863	20.4387	20.7911
60	21.1434	21.4958	21.8482	22.2006	22.5530	22.9054	23.2578	23.6102	23.9626	24.3150
70	24.6673	25.0197	25.3721	25.7245	26.0769	26.4293	26.7817	27.1341	27.4865	27.8389
80	28.1913	28.5436	28.8960	29.2484	29.6008	29.9532	30.3056	30.6680	31.0104	31.3628
90	31.7152	32.0676	32.4199	32.7723	33.1247	33.4771	33.8295	34.1819	34.5343	34.8867

Example: 4567 US bushels = 1585.7582 + 23.6102 hl = 1609.3684 hl

Table 40 HECTOLITRES TO US BUSHELS

1 hl = 2.837 759 32 US bushels

hl	0	100	200	300	400	500	600	700	800	900
	—	283.776	567.552	851.328	1135.104	1418.880	1702.656	1986.432	2270.207	2553.983
	2837.759	3121.535	3405.311	3689.087	3972.863	4256.639	4540.415	4824.191	5107.967	5391.743
	5675.519	5959.295	6243.071	6526.846	6810.622	7094.398	7378.174	7661.950	7945.726	8229.502

hl	0	1	2	3	4	5	6	7	8	9
	—	2.838	5.676	8.513	11.351	14.189	17.027	19.864	22.702	25.540
	28.378	31.215	34.053	36.891	39.729	42.566	45.404	48.242	51.080	53.917
	56.755	59.593	62.431	65.268	68.106	70.944	73.782	76.620	79.457	82.295
	85.133	87.971	90.808	93.646	96.484	99.322	102.159	104.997	107.835	110.673
	113.510	116.348	119.186	122.024	124.861	127.699	130.537	133.375	136.212	139.050
	141.888	144.726	147.563	150.401	153.239	156.077	158.915	161.752	164.590	167.428
	170.266	173.103	175.941	178.779	181.617	184.454	187.292	190.130	192.968	195.805
	198.643	201.481	204.319	207.156	209.994	212.832	215.670	218.507	221.345	224.183
	227.021	229.859	232.696	235.534	238.372	241.210	244.047	246.885	249.723	252.561
	255.398	258.236	261.074	263.912	266.749	269.587	272.425	275.263	278.100	280.938

Example: 402 hl = 1135.104 + 5.676 US bushels = 1140.780 US bushels

Table 41   UK BUSHELS PER ACRE TO HECTOLITRES PER HECTARE

1 UK bushel per acre $=$ 0.898 691 01 hl ha^{-1}

(a)

UK bushels per acre	0	100	200	300	400	500	600	700	800	900
0	—	89.87	179.74	269.61	359.48	449.35	539.21	629.08	718.95	808.82
1000	898.69	988.56	1078.43	1168.30	1258.17	1348.04	1437.91	1527.77	1617.64	1707.51
2000	1797.38	1887.25	1977.12	2066.99	2156.86	2246.73	2336.60	2426.47	2516.33	2606.20

(b)

UK bushels per acre	0	1	2	3	4	5	6	7	8	9
0	—	0.90	1.80	2.70	3.59	4.49	5.39	6.29	7.19	8.09
10	8.99	9.89	10.78	11.68	12.58	13.48	14.38	15.28	16.18	17.08
20	17.97	18.87	19.77	20.67	21.57	22.47	23.37	24.26	25.16	26.06
30	26.96	27.86	28.76	29.66	30.56	31.45	32.35	33.25	34.15	35.05
40	35.95	36.85	37.75	38.64	39.54	40.44	41.34	42.24	43.14	44.04
50	44.93	45.83	46.73	47.63	48.53	49.43	50.33	51.23	52.12	53.02
60	53.92	54.82	55.72	56.62	57.52	58.41	59.31	60.21	61.11	62.01
70	62.91	63.81	64.71	65.60	66.50	67.40	68.30	69.20	70.10	71.00
80	71.90	72.79	73.69	74.59	75.49	76.39	77.29	78.19	79.08	79.98
90	80.88	81.78	82.68	83.58	84.48	85.38	86.27	87.17	88.07	88.97

Example: 293 UK bushels per acre $=$ 179.74 $+$ 83.58 hl ha^{-1} $=$ 263.32 hl ha^{-1}

Table 42   HECTOLITRES PER HECTARE TO UK BUSHELS PER ACRE

1 hl ha^{-1} $=$ 1.112 729 5 UK bushels per acre

(a)

hectolitres per hectare	0	100	200	300	400	500	600	700	800	900
0	—	111.27	222.55	333.82	445.09	556.36	667.64	778.91	890.18	1001.46
1000	1112.73	1224.00	1335.28	1446.55	1557.82	1669.09	1780.37	1891.64	2002.91	2114.19
2000	2225.46	2336.73	2448.00	2559.28	2670.55	2781.82	2893.10	3004.37	3115.64	3226.92

(b)

hectolitres per hectare	0	1	2	3	4	5	6	7	8	9
0	—	1.11	2.23	3.34	4.45	5.56	6.68	7.79	8.90	10.01
10	11.13	12.24	13.35	14.47	15.58	16.69	17.80	18.92	20.03	21.14
20	22.25	23.37	24.48	25.59	26.71	27.82	28.93	30.04	31.16	32.27
30	33.38	34.49	35.61	36.72	37.83	38.95	40.06	41.17	42.28	43.40
40	44.51	45.62	46.73	47.85	48.96	50.07	51.19	52.30	53.41	54.52
50	55.64	56.75	57.86	58.97	60.09	61.20	62.31	63.43	64.54	65.65
60	66.76	67.88	68.99	70.10	71.21	72.33	73.44	74.55	75.67	76.78
70	77.89	79.00	80.12	81.23	82.34	83.45	84.57	85.68	86.79	87.91
80	89.02	90.13	91.24	92.36	93.47	94.58	95.69	96.81	97.92	99.03
90	100.15	101.26	102.37	103.48	104.60	105.71	106.82	107.93	109.05	110.16

Example: 727 hl ha^{-1} $=$ 778.91 $+$ 30.04 UK bushels per acre $=$ 808.95 UK bushels per acre

Table 43   US BUSHELS PER ACRE TO HECTOLITRES PER HECTARE

1 US bushel per acre = 0.870 776 39 hl ha^{-1}

(a)

US bushels per acre	0	100	200	300	400	500	600	700	800	900
0	—	87.08	174.16	261.23	348.31	435.39	522.47	609.54	696.62	783.70
1000	870.78	957.85	1044.93	1132.01	1219.09	1306.16	1393.24	1480.32	1567.40	1654.48
2000	1741.55	1828.63	1915.71	2002.79	2089.86	2176.94	2264.02	2351.10	2438.17	2525.25

(b)

US bushels per acre	0	1	2	3	4	5	6	7	8	9
0	—	0.87	1.74	2.61	3.48	4.35	5.22	6.10	6.97	7.84
10	8.71	9.58	10.45	11.32	12.19	13.06	13.93	14.80	15.67	16.54
20	17.42	18.29	19.16	20.03	20.90	21.77	22.64	23.51	24.38	25.25
30	26.12	26.99	27.86	28.74	29.61	30.48	31.35	32.22	33.09	33.96
40	34.83	35.70	36.57	37.44	38.31	39.18	40.06	40.93	41.80	42.67
50	43.54	44.41	45.28	46.15	47.02	47.89	48.76	49.63	50.51	51.38
60	52.25	53.12	53.99	54.86	55.73	56.60	57.47	58.34	59.21	60.08
70	60.95	61.83	62.70	63.57	64.44	65.31	66.18	67.05	67.92	68.79
80	69.66	70.53	71.40	72.27	73.15	74.02	74.89	75.76	76.63	77.50
90	78.37	79.24	80.11	80.98	81.85	82.72	83.59	84.47	85.34	86.21

Example: 316 US bushels per acre = 261.23 + 13.93 hl ha^{-1} = 275.16 hl ha^{-1}

Table 44   HECTOLITRES PER HECTARE TO US BUSHELS PER ACRE

1 hl ha^{-1} = 1.148 400 5 US bushels per acre

(a)

hectolitres per hectare	0	100	200	300	400	500	600	700	800	900
0	—	114.84	229.68	344 52	459.36	574.20	689.04	803.88	918.72	1033.56
1000	1148.40	1263.24	1378.08	1492.92	1607.76	1722.60	1837.44	1952.28	2067.12	2181.96
2000	2296.80	2411.64	2526.48	2641.32	2756.16	2871.00	2985.84	3100.68	3215.52	3330.36

(b)

hectolitres per hectare	0	1	2	3	4	5	6	7	8	9
0	—	1.15	2.30	3.45	4.59	5.74	6.89	8.04	9.19	10.34
10	11.48	12.63	13.78	14.93	16.08	17.23	18.37	19.52	20.67	21.82
20	22.97	24.12	25.26	26.41	27.56	28.71	29.86	31.01	32.16	33.30
30	34.45	35.60	36.75	37.90	39.05	40.19	41.34	42.49	43.64	44.79
40	45.94	47.08	48.23	49.38	50.53	51.68	52.83	53.97	55.12	56.27
50	57.42	58.57	59.72	60.87	62.01	63.16	64.31	65.46	66.61	67.76
60	68.90	70.05	71.20	72.35	73.50	74.65	75.79	76.94	78.09	79.24
70	80.39	81.54	82.68	83.83	84.98	86.13	87.28	88.43	89.58	90.72
80	91.87	93.02	94.17	95.32	96.47	97.61	98.76	99.91	101.06	102.21
90	103.36	104.50	105.65	106.80	107.95	109.10	110.25	111.39	112.54	113.69

Example: 1586 hl ha^{-1} = 1722.60 + 98.76 US bushels per acre
= 1821.36 US bushels per acre

Table 45   DEGREES, MINUTES AND SECONDS TO RADIANS

180° = π rad = 3.141 592 653 6 rad   1° = 0.017 453 292 5 rad

(a)

degrees	0	1°	2°	3°	4°	5°	6°	7°	8°	9'
0	—	0.017453	0.034907	0.052360	0.069813	0.087266	0.104720	0.122173	0.139626	0.157080
10	0.174533	0.191986	0.209440	0.226893	0.244346	0.261799	0.279253	0.296706	0.314159	0.331613
20	0.349066	0.366519	0.383972	0.401426	0.418879	0.436332	0.453786	0.471239	0.488692	0.506145
30	0.523599	0.541052	0.558505	0.575959	0.593412	0.610865	0.628319	0.645772	0.663225	0.680678
40	0.698132	0.715585	0.733038	0.750492	0.767945	0.785398	0.802851	0.820305	0.837758	0.855211
50	0.872665	0.890118	0.907571	0.925025	0.942478	0.959931	0.977384	0.994838	1.012291	1.029744
60	1.047198	1.064651	1.082104	1.099557	1.117011	1.134464	1.151917	1.169371	1.186824	1.204277
70	1.221730	1.239184	1.256637	1.274090	1.291544	1.308997	1.326450	1.343904	1.361357	1.378810
80	1.396263	1.413717	1.431170	1.448623	1.466077	1.483530	1.500983	1.518436	1.535890	1.553343
90	1.570796	1.588250	1.605703	1.623156	1.640609	1.658063	1.675516	1.692969	1.710423	1.727876
100	1.745329	1.762783	1.780236	1.797689	1.815142	1.832596	1.850049	1.867502	1.884956	1.902409
110	1.919862	1.937315	1.954769	1.972222	1.989675	2.007129	2.024582	2.042035	2.059489	2.076942
120	2.094395	2.111848	2.129302	2.146755	2.164208	2.181662	2.199115	2.216568	2.234021	2.251475
130	2.268928	2.286381	2.303835	2.321288	2.338741	2.356194	2.373648	2.391101	2.408554	2.426008
140	2.443461	2.460914	2.478368	2.495821	2.513274	2.530727	2.548181	2.565634	2.583087	2.600541
150	2.617994	2.635447	2.652900	2.670354	2.687807	2.705260	2.722714	2.740167	2.757620	2.775074
160	2.792527	2.809980	2.827433	2.844887	2.862340	2.879793	2.897247	2.914700	2.932153	2.949606
170	2.967060	2.984513	3.001966	3.019420	3.036873	3.054326	3.071779	3.089233	3.106686	3.124139
180	3.141593	—	—	—	—	—	—	—	—	—

(b)

minutes	0	1'	2'	3'	4'	5'	6'	7'	8'	9'
0	—	0.000291	0.000582	0.000873	0.001164	0.001454	0.001745	0.002036	0.002327	0.002618
10	0.002909	0.003200	0.003491	0.003782	0.004072	0.004363	0.004654	0.004945	0.005236	0.005527
20	0.005818	0.006109	0.006400	0.006690	0.006981	0.007272	0.007563	0.007854	0.008145	0.008436
30	0.008727	0.009018	0.009308	0.009599	0.009890	0.010181	0.010472	0.010763	0.011054	0.011345
40	0.011636	0.011926	0.012217	0.012508	0.012799	0.013090	0.013381	0.013672	0.013963	0.014254
50	0.014544	0.014835	0.015126	0.015417	0.015708	0.015999	0.016290	0.016581	0.016872	0.017162

(c)

seconds	0	1″	2″	3″	4″	5″	6″	7″	8″	9″
0	—	0.000005	0.000010	0.000015	0.000019	0.000024	0.000029	0.000034	0.000039	0.000044
10	0.000048	0.000053	0.000058	0.000063	0.000068	0.000073	0.000078	0.000082	0.000087	0.000092
20	0.000097	0.000102	0.000107	0.000112	0.000116	0.000121	0.000126	0.000131	0.000136	0.000141
30	0.000145	0.000150	0.000155	0.000160	0.000165	0.000170	0.000175	0.000179	0.000184	0.000189
40	0.000194	0.000199	0.000204	0.000208	0.000213	0.000218	0.000223	0.000228	0.000233	0.000238
50	0.000242	0.000247	0.000252	0.000257	0.000262	0.000267	0.000271	0.000276	0.000281	0.000286

Example: 36° 29′ 15″ = 0.628319 + 0.008436 + 0.000073 rad
= 0.636828 rad

Table 46 RADIANS TO DEGREES, MINUTES AND SECONDS

$$1 \text{ rad} = \frac{180°}{\pi} = 57.295\ 779\ 5° = 57°\ 17'\ 45''$$

(a)

radians	0.00	0.01	0.02	0.03	0.04	0.05	0.06	0.07	0.08	0.09
	° ′ ″	° ′ ″	° ′ ″	° ′ ″	° ′ ″	° ′ ″	° ′ ″	° ′ ″	° ′ ″	° ′ ″
0.0	—	0 34 23	1 8 45	1 43 8	2 17 31	2 51 53	3 26 16	4 0 39	4 35 1	5 9 24
0.1	5 43 46	6 18 9	6 52 32	7 26 54	8 1 17	8 35 40	9 10 2	9 44 25	10 18 48	10 53 10
0.2	11 27 33	12 1 56	12 36 18	13 10 41	13 45 4	14 19 26	14 53 49	15 28 11	16 2 34	16 36 57
0.3	17 11 19	17 45 42	18 20 5	18 54 27	19 28 50	20 3 13	20 37 35	21 11 58	21 46 21	22 20 43
0.4	22 55 6	23 29 29	24 3 51	24 38 14	25 12 37	25 46 59	26 21 22	26 55 44	27 30 7	28 4 30
0.5	28 38 52	29 13 15	29 47 38	30 22 0	30 56 23	31 30 46	32 5 8	32 39 31	33 13 54	33 48 16
0.6	34 22 39	34 57 2	35 31 24	36 5 47	36 40 9	37 14 32	37 48 55	38 23 17	38 57 40	39 32 3
0.7	40 6 25	40 40 48	41 15 11	41 49 33	42 23 56	42 58 19	43 32 41	44 7 4	44 41 27	45 15 49
0.8	45 50 12	46 24 34	46 58 57	47 33 20	48 7 42	48 42 5	49 16 28	49 50 50	50 25 13	50 59 36
0.9	51 33 58	52 8 21	52 42 44	53 17 6	53 51 29	54 25 52	55 0 14	55 34 37	56 9 0	56 43 22
1.0	57 17 45	57 52 7	58 26 30	59 0 53	59 35 15	60 9 38	60 44 1	61 18 23	61 52 46	62 27 9
1.1	63 1 31	63 35 54	64 10 17	64 44 39	65 19 2	65 53 25	66 27 47	67 2 10	67 36 32	68 10 55
1.2	68 45 18	69 19 40	69 54 3	70 28 26	71 2 48	71 37 11	72 11 34	72 45 56	73 20 19	73 54 42
1.3	74 29 4	75 3 27	75 37 50	76 12 12	76 46 35	77 20 57	77 55 20	78 29 43	79 4 5	79 38 28
1.4	80 12 51	80 47 13	81 21 36	81 55 59	82 30 21	83 4 44	83 39 7	84 13 29	84 47 52	85 22 15
1.5	85 56 37	86 31 0	87 5 23	87 39 45	88 14 8	88 48 30	89 22 53	89 57 16	90 31 38	91 6 1
1.6	91 40 24	92 14 46	92 49 9	93 23 32	93 57 54	94 32 17	95 6 40	95 41 2	96 15 25	96 49 48
1.7	97 24 10	97 58 33	98 32 55	99 7 18	99 41 41	100 16 3	100 50 26	101 24 49	101 59 11	102 33 34
1.8	103 7 57	103 42 19	104 16 42	104 51 5	105 25 27	105 59 50	106 34 13	107 8 35	107 42 58	108 17 20
1.9	108 51 43	109 26 6	110 0 28	110 34 51	111 9 14	111 43 36	112 17 59	112 52 22	113 26 44	114 1 7
2.0	114 35 30	115 9 52	115 44 15	116 18 38	116 53 0	117 27 23	118 1 46	118 36 8	119 10 31	119 44 53
2.1	120 19 16	120 53 39	121 28 1	122 2 24	122 36 47	123 11 9	123 45 32	124 19 55	124 54 17	125 28 40
2.2	126 3 3	126 37 25	127 11 48	127 46 11	128 20 33	128 54 56	129 29 18	130 3 41	130 38 4	131 12 26
2.3	131 46 49	132 21 12	132 55 34	133 29 57	134 4 20	134 38 42	135 13 5	135 47 28	136 21 50	136 56 13
2.4	137 30 36	138 4 58	138 39 21	139 13 43	139 48 6	140 22 29	140 56 51	141 31 14	142 5 37	142 39 59
2.5	143 14 22	143 48 45	144 23 7	144 57 30	145 31 53	146 6 15	146 40 38	147 15 1	147 49 23	148 23 46
2.6	148 58 8	149 32 31	150 6 54	150 41 16	151 15 39	151 50 2	152 24 24	152 58 47	153 33 10	154 7 32
2.7	154 41 55	155 16 18	155 50 40	156 25 3	156 59 26	157 33 48	158 8 11	158 42 34	159 16 56	159 51 19
2.8	160 25 41	161 0 4	161 34 27	162 8 49	162 43 12	163 17 35	163 51 57	164 26 20	165 0 43	165 35 5
2.9	166 9 28	166 43 51	167 18 13	167 52 36	168 26 59	169 1 21	169 35 44	170 10 6	170 44 29	171 18 52
3.0	171 53 14	172 27 37	173 2 0	173 36 22	174 10 45	174 45 8	175 19 30	175 53 53	176 28 16	177 2 38
3.1	177 37 1	178 11 24	178 45 46	179 20 9	179 54 31	180 28 54	181 3 17	181 37 39	182 12 2	182 46 25

(b)

radians	0.000	0.001	0.002	0.003	0.004	0.005	0.006	0.007	0.008	0.009
	° ′ ″	° ′ ″	° ′ ″	° ′ ″	° ′ ″	° ′ ″	° ′ ″	° ′ ″	° ′ ″	° ′ ″
0.000	—	0 3 26	0 6 53	0 10 19	0 13 45	0 17 11	0 20 38	0 24 4	0 27 30	0 30 56

Example: 2.618 rad = 149° 32′ 31″ + 0° 27′ 30″ = 150° 0′ 1″

Table 47 FEET PER SECOND TO MILES PER HOUR
[ft s^{-1} to mile h^{-1}]   1 ft s^{-1} = 0.681 818 181 8 mile h^{-1}

(a)

feet per second	0	100	200	300	400	500	600	700	800	900
0	—	68.1818	136.3636	204.5455	272.7273	340.9091	409.0909	477.2727	545.4545	613.6364

(b)

feet per second	0	1	2	3	4	5	6	7	8	9
0	—	0.6818	1.3636	2.0455	2.7273	3.4091	4.0909	4.7727	5.4545	6.1364
10	6.8182	7.5000	8.1818	8.8636	9.5455	10.2273	10.9091	11.5909	12.2727	12.9545
20	13.6364	14.3182	15.0000	15.6818	16.3636	17.0455	17.7273	18.4091	19.0909	19.7727
30	20.4545	21.1364	21.8182	22.5000	23.1818	23.8636	24.5455	25.2273	25.9091	26.5909
40	27.2727	27.9545	28.6364	29.3182	30.0000	30.6818	31.3636	32.0455	32.7273	33.4091
50	34.0909	34.7727	35.4545	36.1364	36.8182	37.5000	38.1818	38.8636	39.5455	40.2273
60	40.9091	41.5909	42.2727	42.9545	43.6364	44.3182	45.0000	45.6818	46.3636	47.0455
70	47.7273	48.4091	49.0909	49.7727	50.4545	51.1364	51.8182	52.5000	53.1818	53.8636
80	54.5455	55.2273	55.9091	56.5909	57.2727	57.9545	58.6364	59.3182	60.0000	60.6818
90	61.3636	62.0455	62.7273	63.4091	64.0909	64.7727	65.4545	66.1364	66.8182	67.5000

Example: 182 ft s^{-1} = 68.1818 + 55.9091 mile h^{-1} = 124.0909 mile h^{-1}

Table 48 MILES PER HOUR TO FEET PER SECOND
[mile h^{-1} to ft s^{-1}]   1 mile h^{-1} = 1.466 666 67 ft s^{-1}

(a)

miles per hour	0	100	200	300	400	500	600	700	800	900
0	—	146.6667	293.3333	440.0000	586.6667	733.3333	880.0000	1026.6667	1173.3333	1320.0000

(b)

miles per hour	0	1	2	3	4	5	6	7	8	9
0	—	1.4667	2.9333	4.4000	5.8667	7.3333	8.8000	10.2667	11.7333	13.2000
10	14.6667	16.1333	17.6000	19.0667	20.5333	22.0000	23.4667	24.9333	26.4000	27.8667
20	29.3333	30.8000	32.2667	33.7333	35.2000	36.6667	38.1333	39.6000	41.0667	42.5333
30	44.0000	45.4667	46.9333	48.4000	49.8667	51.3333	52.8000	54.2667	55.7333	57.2000
40	58.6667	60.1333	61.6000	63.0667	64.5333	66.0000	67.4667	68.9333	70.4000	71.8667
50	73.3333	74.8000	76.2667	77.7333	79.2000	80.6667	82.1333	83.6000	85.0667	86.5333
60	88.0000	89.4667	90.9333	92.4000	93.8667	95.3333	96.8000	98.2667	99.7333	101.2000
70	102.6667	104.1333	105.6000	107.0667	108.5333	110.0000	111.4667	112.9333	114.4000	115.8667
80	117.3333	118.8000	120.2667	121.7333	123.2000	124.6667	126.1333	127.6000	129.0667	130.5333
90	132.0000	133.4667	134.9333	136.4000	137.8667	139.3333	140.8000	142.2667	143.7333	145.2000

Example: 121 mile h^{-1} = 146.6667 + 30.8000 ft s^{-1} = 177.4667 ft s^{-1}

Table 49 FEET PER SECOND TO KILOMETRES PER HOUR

[ft s⁻¹ to km h⁻¹]   1 ft s⁻¹ = 1.097 28 km h⁻¹ (exactly)

(a) All values in Table 49(a) are exact

feet per second	0	100	200	300	400	500	600	700	800	900
0	—	109.728	219.456	329.184	438.912	548.640	658.368	768.096	877.824	987.552

(b)

feet per second	0	1	2	3	4	5	6	7	8	9
0	—	1.0973	2.1946	3.2918	4.3891	5.4864	6.5837	7.6810	8.7782	9.8755
10	10.9728	12.0701	13.1674	14.2646	15.3619	16.4592	17.5565	18.6538	19.7510	20.8483
20	21.9456	23.0429	24.1402	25.2374	26.3347	27.4320	28.5293	29.6266	30.7238	31.8211
30	32.9184	34.0157	35.1130	36.2102	37.3075	38.4048	39.5021	40.5994	41.6966	42.7939
40	43.8912	44.9885	46.0858	47.1830	48.2803	49.3776	50.4749	51.5722	52.6694	53.7667
50	54.8640	55.9613	57.0586	58.1558	59.2531	60.3504	61.4477	62.5450	63.6422	64.7395
60	65.8368	66.9341	68.0314	69.1286	70.2259	71.3232	72.4205	73.5178	74.6150	75.7123
70	76.8096	77.9069	79.0042	80.1014	81.1987	82.2960	83.3933	84.4906	85.5878	86.6851
80	87.7824	88.8797	89.9770	91.0742	92.1715	93.2688	94.3661	95.4634	96.5606	97.6579
90	98.7552	99.8525	100.9498	102.0470	103.1443	104.2416	105.3389	106.4362	107.5334	108.6307

Example: 981 ft s⁻¹ = 987.552 + 88.8797 km h⁻¹ = 1076.4317 km h⁻¹

Table 50 KILOMETRES PER HOUR TO FEET PER SECOND

[km h⁻¹ to ft s⁻¹]   1 km h⁻¹ = 0.911 344 415 3 ft s⁻¹

(a)

kilometres per hour	0	100	200	300	400	500	600	700	800	900
0	—	91.1344	182.2689	273.4033	364.5378	455.6722	546.8066	637.9411	729.0755	820.2100

(b)

kilometres per hour	0	1	2	3	4	5	6	7	8	9
0	—	0.9113	1.8227	2.7340	3.6454	4.5567	5.4681	6.3794	7.2908	8.2021
10	9.1134	10.0248	10.9361	11.8475	12.7588	13.6702	14.5815	15.4929	16.4042	17.3155
20	18.2269	19.1382	20.0496	20.9609	21.8723	22.7836	23.6950	24.6063	25.5176	26.4290
30	27.3403	28.2517	29.1630	30.0744	30.9857	31.8971	32.8084	33.7197	34.6311	35.5424
40	36.4538	37.3651	38.2765	39.1878	40.0992	41.0105	41.9218	42.8332	43.7445	44.6559
50	45.5672	46.4786	47.3899	48.3013	49.2126	50.1239	51.0353	51.9466	52.8580	53.7693
60	54.6807	55.5920	56.5034	57.4147	58.3260	59.2374	60.1487	61.0601	61.9714	62.8828
70	63.7941	64.7055	65.6168	66.5281	67.4395	68.3508	69.2622	70.1735	71.0849	71.9962
80	72.9076	73.8189	74.7302	75.6416	76.5529	77.4643	78.3756	79.2870	80.1983	81.1097
90	82.0210	82.9323	83.8437	84.7550	85.6664	86.5777	87.4891	88.4004	89.3118	90.2231

Example: 643 km h⁻¹ = 546.8066 + 39.1878 ft s⁻¹ = 585.9944 ft s⁻¹

Table 51 MILES PER HOUR TO METRES PER SECOND

[mile h^{-1} to m s^{-1}]   1 mile h^{-1} = 0.447 04 m s^{-1} (exactly)
All values in Table 51(a) are exact

(a)

miles per hour	0	100	200	300	400	500	600	700	800	900
0	—	44.704	89.408	134.112	178.816	223.520	268.224	312.928	357.632	402.336

(b)

miles per hour	0	1	2	3	4	5	6	7	8	9
0	—	0.4470	0.8941	1.3411	1.7882	2.2352	2.6822	3.1293	3.5763	4.0234
10	4.4704	4.9174	5.3645	5.8115	6.2586	6.7056	7.1526	7.5997	8.0467	8.4938
20	8.9408	9.3878	9.8349	10.2819	10.7290	11.1760	11.6230	12.0701	12.5171	12.9642
30	13.4112	13.8582	14.3053	14.7523	15.1994	15.6464	16.0934	16.5405	16.9875	17.4346
40	17.8816	18.3286	18.7757	19.2227	19.6698	20.1168	20.5638	21.0109	21.4579	21.9050
50	22.3520	22.7990	23.2461	23.6931	24.1402	24.5872	25.0342	25.4813	25.9283	26.3754
60	26.8224	27.2694	27.7165	28.1635	28.6106	29.0576	29.5046	29.9517	30.3987	30.8458
70	31.2928	31.7398	32.1869	32.6339	33.0810	33.5280	33.9750	34.4221	34.8691	35.3162
80	35.7632	36.2102	36.6573	37.1043	37.5514	37.9984	38.4454	38.8925	39.3395	39.7866
90	40.2336	40.6806	41.1277	41.5747	42.0218	42.4688	42.9158	43.3629	43.8099	44.2570

Example: 134 miles h^{-1} = 44.7040 + 15.1994 m s^{-1} = 59.9034 m s^{-1}

Table 52 METRES PER SECOND TO MILES PER HOUR

[m s^{-1} to mile h^{-1}]   1 m s^{-1} = 2.236 936 3 miles h^{-1}

(a)

metres per second	0	100	200	300	400	500	600	700	800	900
0	—	223.6936	447.3873	671.0809	894.7745	1118.4682	1342.1618	1565.8554	1789.5490	2013.2427

(b)

metres per second	0	1	2	3	4	5	6	7	8	9
0	—	2.2369	4.4739	6.7108	8.9477	11.1847	13.4216	15.6586	17.8955	20.1324
10	22.3694	24.6063	26.8432	29.0802	31.3171	33.5540	35.7910	38.0279	40.2649	42.5018
20	44.7387	46.9757	49.2126	51.4495	53.6865	55.9234	58.1603	60.3973	62.6342	64.8712
30	67.1081	69.3450	71.5820	73.8189	76.0558	78.2928	80.5297	82.7666	85.0036	87.2405
40	89.4775	91.7144	93.9513	96.1883	98.4252	100.6621	102.8991	105.1360	107.3729	109.6099
50	111.8468	114.0838	116.3207	118.5576	120.7946	123.0315	125.2684	127.5054	129.7423	131.9792
60	134.2162	136.4531	138.6901	140.9270	143.1639	145.4009	147.6378	149.8747	152.1117	154.3486
70	156.5855	158.8225	161.0594	163.2963	165.5333	167.7702	170.0072	172.2441	174.4810	176.7180
80	178.9549	181.1918	183.4288	185.6657	187.9026	190.1396	192.3765	194.6135	196.8504	199.0873
90	201.3243	203.5612	205.7981	208.0351	210.2720	212.5089	214.7459	216.9828	219.2198	221.4567

Example: 374 m s^{-1} = 671.0809 + 165.5333 miles h^{-1} = 836.6142 miles h^{-1}

Table 53 MILES PER HOUR TO UK KNOTS

[miles h^{-1} to UK knots]   1 mile h^{-1} = 0.868 421 05 UK knots

(a)

miles per hour	0	100	200	300	400	500	600	700	800	900
0	—	86.8421	173.6842	260.5263	347.3684	434.2105	521.0526	607.8947	694.7368	781.5789

(b)

miles per hour	0	1	2	3	4	5	6	7	8	9
0	—	0.8684	1.7368	2.6053	3.4737	4.3421	5.2105	6.0789	6.9474	7.8158
10	8.6842	9.5526	10.4211	11.2895	12.1579	13.0263	13.8947	14.7632	15.6316	16.5000
20	17.3684	18.2368	19.1053	19.9737	20.8421	21.7105	22.5789	23.4474	24.3158	25.1842
30	26.0526	26.9211	27.7895	28.6579	29.5263	30.3947	31.2632	32.1316	33.0000	33.8684
40	34.7368	35.6053	36.4737	37.3421	38.2105	39.0789	39.9474	40.8158	41.6842	42.5526
50	43.4211	44.2895	45.1579	46.0263	46.8947	47.7632	48.6316	49.5000	50.3684	51.2368
60	52.1053	52.9737	53.8421	54.7105	55.5789	56.4474	57.3158	58.1842	59.0526	59.9211
70	60.7895	61.6579	62.5263	63.3947	64.2632	65.1316	66.0000	66.8684	67.7368	68.6053
80	69.4737	70.3421	71.2105	72.0789	72.9474	73.8158	74.6842	75.5526	76.4211	77.2895
90	78.1579	79.0263	79.8947	80.7632	81.6316	82.5000	83.3684	84.2368	85.1053	85.9737

Example: 624 miles h^{-1} = 521.0526 + 20.8421 UK knots = 541.8947 UK knots

Table 54 UK KNOTS TO MILES PER HOUR

[UK knots to miles h^{-1}]   1 UK knot = 1.151 515 152 miles h^{-1}

(a)

UK knots	0	100	200	300	400	500	600	700	800	900
0	—	115.1515	230.3030	345.4545	460.6061	575.7576	690.9091	806.0606	921.2121	1036.3636

(b)

UK knots	0	1	2	3	4	5	6	7	8	9
0	—	1.1515	2.3030	3.4545	4.6061	5.7576	6.9091	8.0606	9.2121	10.3636
10	11.5152	12.6667	13.8182	14.9697	16.1212	17.2727	18.4242	19.5758	20.7273	21.8788
20	23.0303	24.1818	25.3333	26.4848	27.6364	28.7879	29.9394	31.0909	32.2424	33.3939
30	34.5455	35.6970	36.8485	38.0000	39.1515	40.3030	41.4545	42.6061	43.7576	44.9091
40	46.0606	47.2121	48.3636	49.5152	50.6667	51.8182	52.9697	54.1212	55.2727	56.4242
50	57.5758	58.7273	59.8788	61.0303	62.1818	63.3333	64.4848	65.6364	66.7879	67.9394
60	69.0909	70.2424	71.3939	72.5455	73.6970	74.8485	76.0000	77.1515	78.3030	79.4545
70	80.6061	81.7576	82.9091	84.0606	85.2121	86.3636	87.5152	88.6667	89.8182	90.9697
80	92.1212	93.2727	94.4242	95.5758	96.7273	97.8788	99.0303	100.1818	101.3333	102.4848
90	103.6364	104.7879	105.9394	107.0909	108.2424	109.3939	110.5455	111.6970	112.8485	114.0000

Example: 237 UK knots = 230.3030 + 42.6061 miles h^{-1} = 272.9091 miles h^{-1}

Note: To convert miles per hour to kilometres per hour, refer to Table 9, for example, 30 miles = 48.280 km

hence 30 miles per hour = 48.280 kilometres per hour

Table 55 OUNCES (AVOIRDUPOIS) TO GRAMS
[oz to g]   1 oz = 28.349 523 1 g

(a)

ounces	0	100	200	300	400	500	600	700	800	900
0	—	2834.952	5669.905	8504.857	11339.809	14174.762	17009.714	19844.666	22679.618	25514.571

(b)

ounces	0	1	2	3	4	5	6	7	8	9
0	—	28.350	56.699	85.049	113.398	141.748	170.097	198.447	226.796	255.146
10	283.495	311.845	340.194	368.544	396.893	425.243	453.592	481.942	510.291	538.641
20	566.990	595.340	623.690	652.039	680.389	708.738	737.088	765.437	793.787	822.136
30	850.486	878.835	907.185	935.534	963.884	992.233	1020.583	1048.932	1077.282	1105.631
40	1133.981	1162.330	1190.680	1219.029	1247.379	1275.729	1304.078	1332.428	1360.777	1389.127
50	1417.476	1445.826	1474.175	1502.525	1530.874	1559.224	1587.573	1615.923	1644.272	1672.622
60	1700.971	1729.321	1757.670	1786.020	1814.369	1842.719	1871.069	1899.418	1927.768	1956.117
70	1984.467	2012.816	2041.166	2069.515	2097.865	2126.214	2154.564	2182.913	2211.263	2239.612
80	2267.962	2296.311	2324.661	2353.010	2381.360	2409.709	2438.059	2466.409	2494.758	2523.108
90	2551.457	2579.807	2608.156	2636.506	2664.855	2693.205	2721.554	2749.904	2778.253	2806.603

Example: 343 oz = 8504.857 + 1219.029 g = 9723.886 g

Table 56 GRAMS TO OUNCES (AVOIRDUPOIS)
[g to oz]   1 g = 0.035 273 962 oz

(a)

grams	0	100	200	300	400	500	600	700	800	900
0	—	3.52740	7.05479	10.58219	14.10958	17.63698	21.16438	24.69177	28.21917	31.74657

(b)

grams	0	1	2	3	4	5	6	7	8	9
0	—	0.03527	0.07055	0.10582	0.14110	0.17637	0.21164	0.24692	0.28219	0.31747
10	0.35274	0.38801	0.42329	0.45856	0.49384	0.52911	0.56438	0.59966	0.63493	0.67021
20	0.70548	0.74075	0.77603	0.81130	0.84658	0.88185	0.91712	0.95240	0.98767	1.02294
30	1.05822	1.09349	1.12877	1.16404	1.19931	1.23459	1.26986	1.30514	1.34041	1.37568
40	1.41096	1.44623	1.48151	1.51678	1.55205	1.58733	1.62260	1.65788	1.69315	1.72842
50	1.76370	1.79897	1.83425	1.86952	1.90479	1.94007	1.97534	2.01062	2.04589	2.08116
60	2.11644	2.15171	2.18699	2.22226	2.25753	2.29281	2.32808	2.36336	2.39863	2.43390
70	2.46918	2.50445	2.53973	2.57500	2.61027	2.64555	2.68082	2.71610	2.75137	2.78664
80	2.82192	2.85719	2.89246	2.92774	2.96301	2.99829	3.03356	3.06883	3.10411	3.13938
90	3.17466	3.20993	3.24520	3.28048	3.31575	3.35103	3.38630	3.42157	3.45685	3.49212

Example: 115 g = 3.52740 + 0.52911 oz = 4.05651 oz

Table 57 POUNDS TO KILOGRAMS
[lb to kg]   1 lb = 0.453 592 37 kg (exactly)

(a)

pounds	0	100	200	300	400	500	600	700	800	900
0	—	45.3592	90.7185	136.0777	181.4369	226.7962	272.1554	317.5147	362.8739	408.2331
1000	453.5924	498.9516	544.3108	589.6701	635.0293	680.3886	725.7478	771.1070	816.4663	861.8255
2000	907.1847	952.5440	997.9032	1043.2625	1088.6217	1133.9809	1179.3402	1224.6994	1270.0586	1315.4179

(b)

pounds	0	1	2	3	4	5	6	7	8	9
0	—	0.4536	0.9072	1.3608	1.8144	2.2680	2.7216	3.1751	3.6287	4.0823
10	4.5359	4.9895	5.4431	5.8967	6.3503	6.8039	7.2575	7.7111	8.1647	8.6183
20	9.0718	9.5254	9.9790	10.4326	10.8862	11.3398	11.7934	12.2470	12.7006	13.1542
30	13.6078	14.0614	14.5150	14.9685	15.4221	15.8757	16.3293	16.7829	17.2365	17.6901
40	18.1437	18.5973	19.0509	19.5045	19.9581	20.4117	20.8652	21.3188	21.7724	22.2260
50	22.6796	23.1332	23.5868	24.0404	24.4940	24.9476	25.4012	25.8548	26.3084	26.7619
60	27.2155	27.6691	28.1227	28.5763	29.0299	29.4835	29.9371	30.3907	30.8443	31.2979
70	31.7515	32.2051	32.6587	33.1122	33.5658	34.0194	34.4730	34.9266	35.3802	35.8338
80	36.2874	36.7410	37.1946	37.6482	38.1018	38.5554	39.0089	39.4625	39.9161	40.3697
90	40.8233	41.2769	41.7305	42.1841	42.6377	43.0913	43.5449	43.9985	44.4521	44.9056

Example: 2240 lb = 997.9032 + 18.1437 kg = 1016.0469 kg
Note: To convert kilograms to tonnes, shift the decimal point three places to the left

Table 58 KILOGRAMS TO POUNDS
[kg to lb]   1 kg = 2.204 622 62 lb

(a)

kilo-grams	0	100	200	300	400	500	600	700	800	900
0	—	220.462	440.925	661.387	881.849	1102.311	1322.774	1543.236	1763.698	1984.160
1000	2204.623	2425.085	2645.547	2866.009	3086.472	3306.934	3527.396	3747.858	3968.321	4188.783
2000	4409.245	4629.708	4850.170	5070.632	5291.094	5511.557	5732.019	5952.481	6172.943	6393.406

(b)

kilo-grams	0	1	2	3	4	5	6	7	8	9
0	—	2.205	4.409	6.614	8.818	11.023	13.228	15.432	17.637	19.842
10	22.046	24.251	26.455	28.660	30.865	33.069	35.274	37.479	39.683	41.888
20	44.092	46.297	48.502	50.706	52.911	55.116	57.320	59.525	61.729	63.934
30	66.139	68.343	70.548	72.753	74.957	77.162	79.366	81.571	83.776	85.980
40	88.185	90.390	92.594	94.799	97.003	99.208	101.413	103.617	105.822	108.027
50	110.231	112.436	114.640	116.845	119.050	121.254	123.459	125.663	127.868	130.073
60	132.277	134.482	136.687	138.891	141.096	143.300	145.505	147.710	149.914	152.119
70	154.324	156.528	158.733	160.937	163.142	165.347	167.551	169.756	171.961	174.165
80	176.370	178.574	180.779	182.984	185.188	187.393	189.598	191.802	194.007	196.211
90	198.416	200.621	202.825	205.030	207.235	209.439	211.644	213.848	216.053	218.258

Example: 984 kg = 1984.160 + 185.188 lb = 2169.348 lb
Note: The above tables can also be used to determine Mass Rate of Flow, e.g. Table 57, pounds per second to kilograms per second, and Table 58, kilograms per second to pounds per second

Table 59 UK TONS TO TONNES
1 UK ton = 1.016 046 908 8 t (exactly)

(a)

UK tons	0	100	200	300	400	500	600	700	800	900
0	—	101.6047	203.2094	304.8141	406.4188	508.0235	609.6281	711.2328	812.8375	914.4422
1000	1016.0469	1117.6516	1219.2563	1320.8610	1422.4657	1524.0704	1625.6751	1727.2797	1828.8844	1930.4891

(b)

UK tons	0	1	2	3	4	5	6	7	8	9
0	—	1.0160	2.0321	3.0481	4.0642	5.0802	6.0963	7.1123	8.1284	9.1444
10	10.1605	11.1765	12.1926	13.2086	14.2247	15.2407	16.2568	17.2728	18.2888	19.3049
20	20.3209	21.3370	22.3530	23.3691	24.3851	25.4012	26.4172	27.4333	28.4493	29.4654
30	30.4814	31.4975	32.5135	33.5295	34.5456	35.5616	36.5777	37.5937	38.6098	39.6258
40	40.6419	41.6579	42.6740	43.6900	44.7061	45.7221	46.7382	47.7542	48.7703	49.7863
50	50.8023	51.8184	52.8344	53.8505	54.8665	55.8826	56.8986	57.9147	58.9307	59.9468
60	60.9628	61.9789	62.9949	64.0110	65.0270	66.0430	67.0591	68.0751	69.0912	70.1072
70	71.1233	72.1393	73.1554	74.1714	75.1875	76.2035	77.2196	78.2356	79.2517	80.2677
80	81.2838	82.2998	83.3158	84.3319	85.3479	86.3640	87.3800	88.3961	89.4121	90.4282
90	91.4442	92.4603	93.4763	94.4924	95.5084	96.5245	97.5405	98.5566	99.5726	100.5886

Example: 1658 UK tons = 1625.6751 + 58.9307 t = 1684.6058 t

Table 60 TONNES TO UK TONS
1 t = 0.984 206 528 UK ton

(a)

tonnes	0	100	200	300	400	500	600	700	800	900
0	—	98.4207	196.8413	295.2620	393.6826	492.1033	590.5239	688.9446	787.3652	885.7859
1000	984.2065	1082.6272	1181.0478	1279.4685	1377.8891	1476.3098	1574.7304	1673.1511	1771.5718	1869.9924

(b)

tonnes	0	1	2	3	4	5	6	7	8	9
0	—	0.9842	1.9684	2.9526	3.9368	4.9210	5.9052	6.8894	7.8737	8.8579
10	9.8421	10.8263	11.8105	12.7947	13.7789	14.7631	15.7473	16.7315	17.7157	18.6999
20	19.6841	20.6683	21.6525	22.6368	23.6210	24.6052	25.5894	26.5736	27.5578	28.5420
30	29.5262	30.5104	31.4946	32.4788	33.4630	34.4472	35.4314	36.4156	37.3998	38.3841
40	39.3683	40.3525	41.3367	42.3209	43.3051	44.2893	45.2735	46.2577	47.2419	48.2261
50	49.2103	50.1945	51.1787	52.1629	53.1472	54.1314	55.1156	56.0998	57.0840	58.0682
60	59.0524	60.0366	61.0208	62.0050	62.9892	63.9734	64.9576	65.9418	66.9260	67.9103
70	68.8945	69.8787	70.8629	71.8471	72.8313	73.8155	74.7997	75.7839	76.7681	77.7523
80	78.7365	79.7207	80.7049	81.6891	82.6733	83.6576	84.6418	85.6260	86.6102	87.5944
90	88.5786	89.5628	90.5470	91.5312	92.5154	93.4996	94.4838	95.4680	96.4522	97.4364

Example: 1826 t = 1771.5718 + 25.5894 UK tons = 1797.1612 UK tons
Note: Table 59 can also be used for converting UK tons per hour to tonnes per hour, and Table 60 for converting tonnes per hour to UK tons per hour.

Table 61 US TONS TO TONNES

1 US ton = 0.907 184 74 t (exactly)

(a)

US tons	0	100	200	300	400	500	600	700	800	900
0	—	90.7185	181.4369	272.1554	362.8739	453.5924	544.3108	635.0293	725.7478	816.4663
1000	907.1847	997.9032	1088.6217	1179.3402	1270.0586	1360.7771	1451.4956	1542.2141	1632.9325	1723.6510

(b)

US tons	0	1	2	3	4	5	6	7	8	9
0	—	0.9072	1.8144	2.7216	3.6287	4.5359	5.4431	6.3503	7.2575	8.1647
10	9.0718	9.9790	10.8862	11.7934	12.7006	13.6078	14.5150	15.4221	16.3293	17.2365
20	18.1437	19.0509	19.9581	20.8652	21.7724	22.6796	23.5868	24.4940	25.4012	26.3084
30	27.2155	28.1227	29.0299	29.9371	30.8443	31.7515	32.6587	33.5658	34.4730	35.3802
40	36.2874	37.1946	38.1018	39.0089	39.9161	40.8233	41.7305	42.6377	43.5449	44.4521
50	45.3592	46.2664	47.1736	48.0808	48.9880	49.8952	50.8023	51.7095	52.6167	53.5239
60	54.4311	55.3383	56.2455	57.1526	58.0598	58.9670	59.8742	60.7814	61.6886	62.5957
70	63.5029	64.4101	65.3173	66.2245	67.1317	68.0389	68.9460	69.8532	70.7604	71.6676
80	72.5748	73.4820	74.3891	75.2963	76.2035	77.1107	78.0179	78.9251	79.8323	80.7394
90	81.6466	82.5538	83.4610	84.3682	85.2754	86.1826	87.0897	87.9969	88.9041	89.8113

Example: 1102 US tons = 997.9032 + 1.8144 t = 999.7176 t

Table 62 TONNES TO US TONS

1 t = 1.102 311 31 US tons

(a)

tonnes	0	100	200	300	400	500	600	700	800	900
0	—	110.2311	220.4623	330.6934	440.9245	551.1557	661.3868	771.6179	881.8490	992.0802
1000	1102.3113	1212.5424	1322.7736	1433.0047	1543.2358	1653.4670	1763.6981	1873.9292	1984.1604	2094.3915

(b)

tonnes	0	1	2	3	4	5	6	7	8	9
0	—	1.1023	2.2046	3.3069	4.4092	5.5116	6.6139	7.7162	8.8185	9.9208
10	11.0231	12.1254	13.2277	14.3300	15.4324	16.5347	17.6370	18.7393	19.8416	20.9439
20	22.0462	23.1485	24.2508	25.3532	26.4555	27.5578	28.6601	29.7624	30.8647	31.9670
30	33.0693	34.1717	35.2740	36.3763	37.4786	38.5809	39.6832	40.7855	41.8878	42.9901
40	44.0925	45.1948	46.2971	47.3994	48.5017	49.6040	50.7063	51.8086	52.9109	54.0133
50	55.1156	56.2179	57.3202	58.4225	59.5248	60.6271	61.7294	62.8317	63.9341	65.0364
60	66.1387	67.2410	68.3433	69.4456	70.5479	71.6502	72.7525	73.8549	74.9572	76.0595
70	77.1618	78.2641	79.3664	80.4687	81.5710	82.6733	83.7757	84.8780	85.9803	87.0826
80	88.1849	89.2872	90.3895	91.4918	92.5942	93.6965	94.7988	95.9011	97.0034	98.1057
90	99.2080	100.3103	101.4126	102.5150	103.6173	104.7196	105.8219	106.9242	108.0265	109.1288

Example: 907 t = 992.0802 + 7.7162 US tons = 999.7964 US tons
Note: The term 'long' ton is sometimes used to refer to the ton of 2240 lb (UK ton) ; in the U.S.A. the word ton refers to the 'short' ton of 2000 lb (US ton) ; the tonne is often referred to as a 'metric ton' in the U.K. and the U.S.A.

Table 63   UK TONS TO US TONS

1 UK ton = 1.12 US tons (exactly)
All values in Tables 63(a) and (b) are exact.

(a)

UK tons	0	100	200	300	400	500	600	700	800	900
0	—	112.0	224.0	336.0	448.0	560.0	672.0	784.0	896.0	1008.0
1000	1120.0	1232.0	1344.0	1456.0	1568.0	1680.0	1792.0	1904.0	2016.0	2128.0

(b)

UK tons	0	1	2	3	4	5	6	7	8	9
0	—	1.12	2.24	3.36	4.48	5.60	6.72	7.84	8.96	10.08
10	11.20	12.32	13.44	14.56	15.68	16.80	17.92	19.04	20.16	21.28
20	22.40	23.52	24.64	25.76	26.88	28.00	29.12	30.24	31.36	32.48
30	33.60	34.72	35.84	36.96	38.08	39.20	40.32	41.44	42.56	43.68
40	44.80	45.92	47.04	48.16	49.28	50.40	51.52	52.64	53.76	54.88
50	56.00	57.12	58.24	59.36	60.48	61.60	62.72	63.84	64.96	66.08
60	67.20	68.32	69.44	70.56	71.68	72.80	73.92	75.04	76.16	77.28
70	78.40	79.52	80.64	81.76	82.88	84.00	85.12	86.24	87.36	88.48
80	89.60	90.72	91.84	92.96	94.08	95.20	96.32	97.44	98.56	99.68
90	100.80	101.92	103.04	104.16	105.28	106.40	107.52	108.64	109.76	110.88

Example: 1775 UK tons = 1904.0 + 84.00 US tons = 1988 US tons (exactly)

Table 64   US TONS TO UK TONS

1 US ton = 0.892 857 143 UK ton

(a)

US tons	0	100	200	300	400	500	600	700	800	900
0	—	89.2857	178.5714	267.8571	357.1429	446.4286	535.7143	625.0000	714.2857	803.5714
1000	892.8571	982.1429	1071.4286	1160.7143	1250.0000	1339.2857	1428.5714	1517.8571	1607.1429	1696.4286

(b)

US tons	0	1	2	3	4	5	6	7	8	9
0	—	0.8929	1.7857	2.6786	3.5714	4.4643	5.3571	6.2500	7.1429	8.0357
10	8.9286	9.8214	10.7143	11.6071	12.5000	13.3929	14.2857	15.1786	16.0714	16.9643
20	17.8571	18.7500	19.6429	20.5357	21.4286	22.3214	23.2143	24.1071	25.0000	25.8929
30	26.7857	27.6786	28.5714	29.4643	30.3571	31.2500	32.1429	33.0357	33.9286	34.8214
40	35.7143	36.6071	37.5000	38.3929	39.2857	40.1786	41.0714	41.9643	42.8571	43.7500
50	44.6429	45.5357	46.4286	47.3214	48.2143	49.1071	50.0000	50.8929	51.7857	52.6786
60	53.5714	54.4643	55.3571	56.2500	57.1429	58.0357	58.9286	59.8214	60.7143	61.6071
70	62.5000	63.3929	64.2857	65.1786	66.0714	66.9643	67.8571	68.7500	69.6429	70.5357
80	71.4286	72.3214	73.2143	74.1071	75.0000	75.8929	76.7857	77.6786	78.5714	79.4643
90	80.3571	81.2500	82.1429	83.0357	83.9286	84.8214	85.7143	86.6071	87.5000	88.3929

Example: 1892 US tons = 1607.1429 + 82.1429 UK tons = 1689.2858 UK tons

Table 65 POUNDS PER SQUARE INCH TO KILOGRAMS PER SQUARE CENTIMETRE

[lb in^{-2} to kg cm^{-2}] 1 lb in^{-2} = 0.070 306 958 kg cm^{-2}

(a)

lb in^{-2}	0	100	200	300	400	500	600	700	800	900
0	—	7.0307	14.0614	21.0921	28.1228	35.1535	42.1842	49.2149	56.2456	63.2763

(b)

lb in^{-2}	0	1	2	3	4	5	6	7	8	9
0	—	0.0703	0.1406	0.2109	0.2812	0.3515	0.4218	0.4921	0.5625	0.6328
10	0.7031	0.7734	0.8437	0.9140	0.9843	1.0546	1.1249	1.1952	1.2655	1.3358
20	1.4061	1.4764	1.5468	1.6171	1.6874	1.7577	1.8280	1.8983	1.9686	2.0389
30	2.1092	2.1795	2.2498	2.3201	2.3904	2.4607	2.5311	2.6014	2.6717	2.7420
40	2.8123	2.8826	2.9529	3.0232	3.0935	3.1638	3.2341	3.3044	3.3747	3.4450
50	3.5153	3.5857	3.6560	3.7263	3.7966	3.8669	3.9372	4.0075	4.0778	4.1481
60	4.2184	4.2887	4.3590	4.4293	4.4996	4.5700	4.6403	4.7106	4.7809	4.8512
70	4.9215	4.9918	5.0621	5.1324	5.2027	5.2730	5.3433	5.4136	5.4839	5.5542
80	5.6246	5.6949	5.7652	5.8355	5.9058	5.9761	6.0464	6.1167	6.1870	6.2573
90	6.3276	6.3979	6.4682	6.5385	6.6089	6.6792	6.7495	6.8198	6.8901	6.9604

Example: 321 lb in^{-2} = 21.0921 + 1.4764 kg cm^{-2} = 22.5685 kg cm^{-2}

Table 66 KILOGRAMS PER SQUARE CENTIMETRE TO POUNDS PER SQUARE INCH

[kg cm^{-2} to lb in^{-2}] 1 kg cm^{-2} = 14.223 343 lb in^{-2}

(a)

kg cm^{-2}	—	100	200	300	400	500	600	700	800	900
0	—	1422.334	2844.669	4267.003	5689.337	7111.672	8534.006	9956.340	11378.674	12801.009

(b)

kg cm^{-2}	0	1	2	3	4	5	6	7	8	9
0	—	14.223	28.447	42.670	56.893	71.117	85.340	99.563	113.787	128.010
10	142.233	156.457	170.680	184.903	199.127	213.350	227.573	241.797	256.020	270.244
20	284.467	298.690	312.914	327.137	341.360	355.584	369.807	384.030	398.254	412.477
30	426.700	440.924	455.147	469.370	483.594	497.817	512.040	526.264	540.487	554.710
40	568.934	583.157	597.380	611.604	625.827	640.050	654.274	668.497	682.720	696.944
50	711.167	725.390	739.614	753.837	768.061	782.284	796.507	810.731	824.954	839.177
60	853.401	867.624	881.847	896.071	910.294	924.517	938.741	952.964	967.187	981.411
70	995.634	1009.857	1024.081	1038.304	1052.527	1066.751	1080.974	1095.197	1109.421	1123.644
80	1137.867	1152.091	1166.314	1180.537	1194.761	1208.984	1223.207	1237.431	1251.654	1265.878
90	1280.101	1294.324	1308.548	1322.771	1336.994	1351.218	1365.441	1379.664	1393.888	1408.111

Example: 256 kg cm^{-2} = 2844.669 + 796.507 lb in^{-2} = 3641.176 lb in^{-2}
Note: The above tables can also be used to determine Pressure and Stress, e.g. pounds-force per square inch to kilograms-force per square centimetre.

Table 67 CUBIC FEET PER UK TON TO CUBIC METRES PER TONNE

$[ft^3\ ton^{-1}\ to\ m^3\ t^{-1}]$ $1\ ft^3\ ton^{-1} = 0.027\ 869\ 625\ 2\ m^3\ t^{-1}$.

(a)

$ft^3\ ton^{-1}$	0	100	200	300	400	500	600	700	800	900
0	—	2.7870	5.5739	8.3609	11.1479	13.9348	16.7218	19.5087	22.2957	25.0827

(b)

$ft^3\ ton^{-1}$	0	1	2	3	4	5	6	7	8	9
0	—	0.0279	0.0557	0.0836	0.1115	0.1393	0.1672	0.1951	0.2230	0.2508
10	0.2787	0.3066	0.3344	0.3623	0.3902	0.4180	0.4459	0.4738	0.5017	0.5295
20	0.5574	0.5853	0.6131	0.6410	0.6689	0.6967	0.7246	0.7525	0.7803	0.8082
30	0.8361	0.8640	0.8918	0.9197	0.9476	0.9754	1.0033	1.0312	1.0590	1.0869
40	1.1148	1.1427	1.1705	1.1984	1.2263	1.2541	1.2820	1.3099	1.3377	1.3656
50	1.3935	1.4214	1.4492	1.4771	1.5050	1.5328	1.5607	1.5886	1.6164	1.6443
60	1.6722	1.7000	1.7279	1.7558	1.7837	1.8115	1.8394	1.8673	1.8951	1.9230
70	1.9509	1.9787	2.0066	2.0345	2.0624	2.0902	2.1181	2.1460	2.1738	2.2017
80	2.2296	2.2574	2.2853	2.3132	2.3410	2.3689	2.3968	2.4247	2.4525	2.4804
90	2.5083	2.5361	2.5640	2.5919	2.6197	2.6476	2.6755	2.7034	2.7312	2.7591

Example: $379\ ft^3\ ton^{-1} = 8.3609 + 2.2017\ m^3\ t^{-1} = 10.5626\ m^3\ t^{-1}$

Table 68 CUBIC METRES PER TONNE TO CUBIC FEET PER UK TON

$[m^3\ t^{-1}\ to\ ft^3\ ton^{-1}]$ $1\ m^3\ t^{-1} = 35.881\ 358\ 0\ ft^3\ ton^{-1}$

(a)

$m^3\ t^{-1}$	0	100	200	300	400	500	600	700	800	900
0	—	3588.14	7176.27	10764.41	14352.54	17940.68	21528.81	25116.95	28705.09	32293.22

(b)

$m^3\ t^{-1}$	0	1	2	3	4	5	6	7	8	9
0	—	35.88	71.76	107.64	143.53	179.41	215.29	251.17	287.05	322.93
10	358.81	394.69	430.58	466.46	502.34	538.22	574.10	609.98	645.86	681.75
20	717.63	753.51	789.39	825.27	861.15	897.03	932.92	968.80	1004.68	1040.56
30	1076.44	1112.32	1148.20	1184.08	1219.97	1255.85	1291.73	1327.61	1363.49	1399.37
40	1435.25	1471.14	1507.02	1542.90	1578.78	1614.66	1650.54	1686.42	1722.31	1758.19
50	1794.07	1829.95	1865.83	1901.71	1937.59	1973.47	2009.36	2045.24	2081.12	2117.00
60	2152.88	2188.76	2224.64	2260.53	2296.41	2332.29	2368.17	2404.05	2439.93	2475.81
70	2511.70	2547.58	2583.46	2619.34	2655.22	2691.10	2726.98	2762.86	2798.75	2834.63
80	2870.51	2906.39	2942.27	2978.15	3014.03	3049.92	3085.80	3121.68	3157.56	3193.44
90	3229.32	3265.20	3301.08	3336.97	3372.85	3408.73	3444.61	3480.49	3516.37	3552.25

Example: $191\ m^3\ t^{-1} = 3588.14 + 3265.20\ ft^3\ ton^{-1} = 6853.34\ ft^3\ ton^{-1}$

Table 69 CUBIC YARDS PER UK TON TO CUBIC METRES PER TONNE

[$yd^3\ ton^{-1}$ to $m^3\ t^{-1}$] 1 $yd^3\ ton^{-1}$ = 0.752 479 88 $m^3\ t^{-1}$

(a)

$yd^3\ ton^{-1}$	0	100	200	300	400	500	600	700	800	900
0	—	75.2480	150.4960	225.7440	300.9920	376.2399	451.4879	526.7359	601.9839	677.2319

(b)

$yd^3\ ton^{-1}$	0	1	2	3	4	5	6	7	8	9
0	—	0.7525	1.5050	2.2574	3.0099	3.7624	4.5149	5.2674	6.0198	6.7723
10	7.5248	8.2773	9.0298	9.7822	10.5347	11.2872	12.0397	12.7922	13.5446	14.2971
20	15.0496	15.8021	16.5546	17.3070	18.0595	18.8120	19.5645	20.3170	21.0694	21.8219
30	22.5744	23.3269	24.0794	24.8318	25.5843	26.3368	27.0893	27.8418	28.5942	29.3467
40	30.0992	30.8517	31.6042	32.3566	33.1091	33.8616	34.6141	35.3666	36.1190	36.8715
50	37.6240	38.3765	39.1290	39.8814	40.6339	41.3864	42.1389	42.8914	43.6438	44.3963
60	45.1488	45.9013	46.6538	47.4062	48.1587	48.9112	49.6637	50.4162	51.1686	51.9211
70	52.6736	53.4261	54.1786	54.9310	55.6835	56.4360	57.1885	57.9410	58.6934	59.4459
80	60.1984	60.9509	61.7034	62.4558	63.2083	63.9608	64.7133	65.4657	66.2182	66.9707
90	67.7232	68.4757	69.2281	69.9806	70.7331	71.4856	72.2381	72.9905	73.7430	74.4955

Example: 473 $yd^3\ ton^{-1}$ = 300.9920 + 54.9310 $m^3\ t^{-1}$ = 355.9230 $m^3\ t^{-1}$

Table 70 CUBIC METRES PER TONNE TO CUBIC YARDS PER UK TON

[$m^3\ t^{-1}$ to $yd^3\ ton^{-1}$] 1 $m^3\ t^{-1}$ = 1.328 939 2 $yd^3\ ton^{-1}$

(a)

$m^3\ t^{-1}$	0	100	200	300	400	500	600	700	800	900
0	—	132.894	265.788	398.682	531.576	664.470	797.364	930.257	1063.151	1196.045

(b)

$m^3\ t^{-1}$	0	1	2	3	4	5	6	7	8	9
0	—	1.329	2.658	3.987	5.316	6.645	7.974	9.303	10.632	11.960
10	13.289	14.618	15.947	17.276	18.605	19.934	21.263	22.592	23.921	25.250
20	26.579	27.908	29.237	30.566	31.895	33.223	34.552	35.881	37.210	38.539
30	39.868	41.197	42.526	43.855	45.184	46.513	47.842	49.171	50.500	51.829
40	53.158	54.487	55.815	57.144	58.473	59.802	61.131	62.460	63.789	65.118
50	66.447	67.776	69.105	70.434	71.763	73.092	74.421	75.750	77.078	78.407
60	79.736	81.065	82.394	83.723	85.052	86.381	87.710	89.039	90.368	91.697
70	93.026	94.355	95.684	97.013	98.342	99.670	100.999	102.328	103.657	104.986
80	106.315	107.644	108.973	110.302	111.631	112.960	114.289	115.618	116.947	118.276
90	119.605	120.933	122.262	123.591	124.920	126.249	127.578	128.907	130.236	131.565

Example: 365 $m^3\ t^{-1}$ = 398.682 + 86.381 $yd^3\ ton^{-1}$ = 485.063 $yd^3\ ton^{-1}$

Table 71 UK GALLONS PER UK TON TO LITRES PER TONNE

[UKgal ton^{-1} to l t^{-1}]   1 UKgal ton^{-1} = 4.474 293 3 l t^{-1}

(a)

UKgal ton^{-1}	0	100	200	300	400	500	600	700	800	900
0	—	447.43	894.86	1342.29	1789.72	2237.15	2684.58	3132.01	3579.43	4026.86

(b)

UKgal ton^{-1}	0	1	2	3	4	5	6	7	8	9
0	—	4.47	8.95	13.42	17.90	22.37	26.85	31.32	35.79	40.27
10	44.74	49.22	53.69	58.17	62.64	67.11	71.59	76.06	80.54	85.01
20	89.49	93.96	98.43	102.91	107.38	111.86	116.33	120.81	125.28	129.75
30	134.23	138.70	143.18	147.65	152.13	156.60	161.07	165.55	170.02	174.50
40	178.97	183.45	187.92	192.39	196.87	201.34	205.82	210.29	214.77	219.24
50	223.71	228.19	232.66	237.14	241.61	246.09	250.56	255.03	259.51	263.98
60	268.46	272.93	277.41	281.88	286.35	290.83	295.30	299.78	304.25	308.73
70	313.20	317.67	322.15	326.62	331.10	335.57	340.05	344.52	348.99	353.47
80	357.94	362.42	366.89	371.37	375.84	380.31	384.79	389.26	393.74	398.21
90	402.69	407.16	411.63	416.11	420.58	425.06	429.53	434.01	438.48	442.96

Example: 647 UKgal ton^{-1} = 2684.58 + 210.29 l t^{-1} = 2894.87 l t^{-1}

Table 72 LITRES PER TONNE TO UK GALLONS PER UK TON

[l t^{-1} to UKgal ton^{-1}]   1 l t^{-1} = 0.223 499 0 UKgal ton^{-1}

(a)

l t^{-1}	0	100	200	300	400	500	600	700	800	900
0	—	22.350	44.700	67.050	89.400	111.750	134.099	156.449	178.799	201.149

(b)

l t^{-1}	0	1	2	3	4	5	6	7	8	9
0	—	0.223	0.447	0.670	0.894	1.117	1.341	1.564	1.788	2.011
10	2.235	2.458	2.682	2.905	3.129	3.352	3.576	3.799	4.023	4.246
20	4.470	4.693	4.917	5.140	5.364	5.587	5.811	6.034	6.258	6.481
30	6.705	6.928	7.152	7.375	7.599	7.822	8.046	8.269	8.493	8.716
40	8.940	9.163	9.387	9.610	9.834	10.057	10.281	10.504	10.728	10.951
50	11.175	11.398	11.622	11.845	12.069	12.292	12.516	12.739	12.963	13.186
60	13.410	13.633	13.857	14.080	14.304	14.527	14.751	14.974	15.198	15.421
70	15.645	15.868	16.092	16.315	16.539	16.762	16.986	17.209	17.433	17.656
80	17.880	18.103	18.327	18.550	18.774	18.997	19.221	19.444	19.668	19.891
90	20.115	20.338	20.562	20.785	21.009	21.232	21.456	21.679	21.903	22.126

Example: 815 l t^{-1} = 178.799 + 3.352 UKgal ton^{-1} = 182.151 UKgal ton^{-1}

CUBIC FEET PER MINUTE TO CUBIC DECIMETRES PER SECOND

[ft³ min⁻¹ to dm³ s⁻¹]   1 ft³ min⁻¹ = 0.471 947 443 2 dm³ s⁻¹ (exactly)

ft³ min⁻¹	0	100	200	300	400	500	600	700	800	900
0	—	47.1947	94.3895	141.5842	188.7790	235.9737	283.1685	330.3632	377.5580	424.7527

ft³ min⁻¹	0	1	2	3	4	5	6	7	8	9
0	—	0.4719	0.9439	1.4158	1.8878	2.3597	2.8317	3.3036	3.7756	4.2475
10	4.7195	5.1914	5.6634	6.1353	6.6073	7.0792	7.5512	8.0231	8.4951	8.9670
20	9.4389	9.9109	10.3828	10.8548	11.3267	11.7987	12.2706	12.7426	13.2145	13.6865
30	14.1584	14.6304	15.1023	15.5743	16.0462	16.5182	16.9901	17.4621	17.9340	18.4060
40	18.8779	19.3498	19.8218	20.2937	20.7657	21.2376	21.7096	22.1815	22.6535	23.1254
50	23.5974	24.0693	24.5413	25.0132	25.4852	25.9571	26.4291	26.9010	27.3730	27.8449
60	28.3168	28.7888	29.2607	29.7327	30.2046	30.6766	31.1485	31.6205	32.0924	32.5644
70	33.0363	33.5083	33.9802	34.4522	34.9241	35.3961	35.8680	36.3400	36.8119	37.2838
80	37.7558	38.2277	38.6997	39.1716	39.6436	40.1155	40.5875	41.0594	41.5314	42.0033
90	42.4753	42.9472	43.4192	43.8911	44.3631	44.8350	45.3070	45.7789	46.2508	46.7228

Example: 484 ft³ min⁻¹ = 188.7790 + 39.6436 dm³ s⁻¹ = 228.4226 dm³ s⁻¹

CUBIC DECIMETRES PER SECOND TO CUBIC FEET PER MINUTE

[dm³ s⁻¹ to ft³ min⁻¹]   1 dm³ s⁻¹ = 2.118 880 00 ft³ min⁻¹

dm³ s⁻¹	0	100	200	300	400	500	600	700	800	900
0	—	211.8880	423.7760	635.6640	847.5520	1059.4400	1271.3280	1483.2160	1695.1040	1906.9920

dm³ s⁻¹	0	1	2	3	4	5	6	7	8	9
0	—	2.1189	4.2378	6.3566	8.4755	10.5944	12.7133	14.8322	16.9510	19.0699
10	21.1888	23.3077	25.4266	27.5454	29.6643	31.7832	33.9021	36.0210	38.1398	40.2587
20	42.3776	44.4965	46.6154	48.7342	50.8531	52.9720	55.0909	57.2098	59.3286	61.4475
30	63.5664	65.6853	67.8042	69.9230	72.0419	74.1608	76.2797	78.3986	80.5174	82.6363
40	84.7552	86.8741	88.9930	91.1118	93.2307	95.3496	97.4685	99.5874	101.7062	103.8251
50	105.9440	108.0629	110.1818	112.3006	114.4195	116.5384	118.6573	120.7762	122.8950	125.0139
60	127.1328	129.2517	131.3706	133.4894	135.6083	137.7272	139.8461	141.9650	144.0838	146.2027
70	148.3216	150.4405	152.5594	154.6782	156.7971	158.9160	161.0349	163.1538	165.2726	167.3915
80	169.5104	171.6293	173.7482	175.8670	177.9859	180.1048	182.2237	184.3426	186.4614	188.5803
90	190.6992	192.8181	194.9370	197.0558	199.1747	201.2936	203.4125	205.5314	207.6502	209.7691

Example: 271 dm³ s⁻¹ = 423.7760 + 150.4405 ft³ min⁻¹ = 574.2165 ft³ min⁻¹

Table 75 UK GALLONS PER HOUR TO CUBIC METRES PER HOUR

[UKgal h⁻¹ to m³ h⁻¹]   1 UKgal h⁻¹ = 0.004 546 091 878 m³ h⁻¹

(a)

UKgal h⁻¹	0	100	200	300	400	500	600	700	800	900
0	—	0.45461	0.90922	1.36383	1.81844	2.27305	2.72766	3.18226	3.63687	4.09148

(b)

UKgal h⁻¹	0	1	2	3	4	5	6	7	8	9
0	—	0.00455	0.00909	0.01364	0.01818	0.02273	0.02728	0.03182	0.03637	0.04091
10	0.04546	0.05001	0.05455	0.05910	0.06365	0.06819	0.07274	0.07728	0.08183	0.08638
20	0.09092	0.09547	0.10001	0.10456	0.10911	0.11365	0.11820	0.12274	0.12729	0.13184
30	0.13638	0.14093	0.14547	0.15002	0.15457	0.15911	0.16366	0.16821	0.17275	0.17730
40	0.18184	0.18639	0.19094	0.19548	0.20003	0.20457	0.20912	0.21367	0.21821	0.22276
50	0.22730	0.23185	0.23640	0.24094	0.24549	0.25004	0.25458	0.25913	0.26367	0.26822
60	0.27277	0.27731	0.28186	0.28640	0.29095	0.29550	0.30004	0.30459	0.30913	0.31368
70	0.31823	0.32277	0.32732	0.33186	0.33641	0.34096	0.34550	0.35005	0.35460	0.35914
80	0.36369	0.36823	0.37278	0.37733	0.38187	0.38642	0.39096	0.39551	0.40006	0.40460
90	0.40915	0.41369	0.41824	0.42279	0.42733	0.43188	0.43642	0.44097	0.44552	0.45006

Example: 323 UKgal h⁻¹ = 1.36383 + 0.10456 m³ h⁻¹ = 1.46839 m³ h⁻¹

Table 76 CUBIC METRES PER HOUR TO UK GALLONS PER HOUR

[m³ h⁻¹ to UKgal h⁻¹]   1 m³ h⁻¹ = 219.969 157 UKgal h⁻¹

(a)

m³ h⁻¹	0	100	200	300	400	500	600	700	800	900
0	—	21996.9	43993.8	65990.7	87987.7	109984.6	131981.5	153978.4	175975.3	197972.2

(b)

m³ h⁻¹	0	1	2	3	4	5	6	7	8	9
0	—	220.0	439.9	659.9	879.9	1099.8	1319.8	1539.8	1759.8	1979.7
10	2199.7	2419.7	2639.6	2859.6	3079.6	3299.5	3519.5	3739.5	3959.4	4179.4
20	4399.4	4619.4	4839.3	5059.3	5279.3	5499.2	5719.2	5939.2	6159.1	6379.1
30	6599.1	6819.0	7039.0	7259.0	7479.0	7698.9	7918.9	8138.9	8358.8	8578.8
40	8798.8	9018.7	9238.7	9458.7	9678.6	9898.6	10118.6	10338.6	10558.5	10778.5
50	10998.5	11218.4	11438.4	11658.4	11878.3	12098.3	12318.3	12538.2	12758.2	12978.2
60	13198.1	13418.1	13638.1	13858.1	14078.0	14298.0	14518.0	14737.9	14957.9	15177.9
70	15397.8	15617.8	15837.8	16057.7	16277.7	16497.7	16717.7	16937.6	17157.6	17377.6
80	17597.5	17817.5	18037.5	18257.4	18477.4	18697.4	18917.3	19137.3	19357.3	19577.3
90	19797.2	20017.2	20237.2	20457.1	20677.1	20897.1	21117.0	21337.0	21557.0	21776.9

Example: 111 m³ h⁻¹ = 21996.9 + 2419.7 UKgal h⁻¹ = 24416.6 UKgal h⁻¹
Note: Table 75 can also be used for converting UK gallons to cubic metres and Table 76 for converting cubic metres to UK gallo
e.g. 323 UKgal = 1.46839 m³

Table 77 POUNDS PER CUBIC FOOT TO KILOGRAMS PER CUBIC METRE

[lb ft^{-3} to kg m^{-3}]   1 lb ft^{-3} = 16.018 463 374 kg m^{-3}

(a)

lb ft^{-3}	0	100	200	300	400	500	600	700	800	900
0	—	1601.846	3203.693	4805.539	6407.385	8009.232	9611.078	11212.924	12814.771	14416.617

(b)

lb ft^{-3}	0	1	2	3	4	5	6	7	8	9
0	—	16.018	32.037	48.055	64.074	80.092	96.111	112.129	128.148	144.166
10	160.185	176.203	192.222	208.240	224.258	240.277	256.295	272.314	288.332	304.351
20	320.369	336.388	352.406	368.425	384.443	400.462	416.480	432.499	448.517	464.535
30	480.554	496.572	512.591	528.609	544.628	560.646	576.665	592.683	608.702	624.720
40	640.739	656.757	672.775	688.794	704.812	720.831	736.849	752.868	768.886	784.905
50	800.923	816.942	832.960	848.979	864.997	881.015	897.034	913.052	929.071	945.089
60	961.108	977.126	993.145	1009.163	1025.182	1041.200	1057.219	1073.237	1089.256	1105.274
70	1121.292	1137.311	1153.329	1169.348	1185.366	1201.385	1217.403	1233.422	1249.440	1265.459
80	1281.477	1297.496	1313.514	1329.532	1345.551	1361.569	1377.588	1393.606	1409.625	1425.643
90	1441.662	1457.680	1473.699	1489.717	1505.736	1521.754	1537.772	1553.791	1569.809	1585.828

Example: 561 lb ft^{-3} = 8009.232 + 977.126 kg m^{-3} = 8986.358 kg m^{-3}

Table 78 KILOGRAMS PER CUBIC METRE TO POUNDS PER CUBIC FOOT

[kg m^{-3} to lb ft^{-3}]   1 kg m^{-3} = 0.062 427 960 6 lb ft^{-3}

(a)

kg m^{-3}	0	1000	2000	3000	4000	5000	6000	7000	8000	9000
0	—	62.4280	124.8559	187.2839	249.7118	312.1398	374.5678	436.9957	499.4237	561.8516

(b)

kg m^{-3}	0	10	20	30	40	50	60	70	80	90
0	—	0.6243	1.2486	1.8728	2.4971	3.1214	3.7457	4.3700	4.9942	5.6185
100	6.2428	6.8671	7.4914	8.1156	8.7399	9.3642	9.9885	10.6128	11.2370	11.8613
200	12.4856	13.1099	13.7342	14.3584	14.9827	15.6070	16.2313	16.8555	17.4798	18.1041
300	18.7284	19.3527	19.9769	20.6012	21.2255	21.8498	22.4741	23.0983	23.7226	24.3469
400	24.9712	25.5955	26.2197	26.8440	27.4683	28.0926	28.7169	29.3411	29.9654	30.5897
500	31.2140	31.8383	32.4625	33.0868	33.7111	34.3354	34.9597	35.5839	36.2082	36.8325
600	37.4568	38.0811	38.7053	39.3296	39.9539	40.5782	41.2025	41.8267	42.4510	43.0753
700	43.6996	44.3239	44.9481	45.5724	46.1967	46.8210	47.4453	48.0695	48.6938	49.3181
800	49.9424	50.5666	51.1909	51.8152	52.4395	53.0638	53.6880	54.3123	54.9366	55.5609
900	56.1852	56.8094	57.4337	58.0580	58.6823	59.3066	59.9308	60.5551	61.1794	61.8037

Example: 8760 kg m^{-3} = 499.4237 + 47.4453 lb ft^{-3} = 546.8690 lb ft^{-3}

Table 79 POUNDS PER UK GALLON TO KILOGRAMS PER LITRE
[lb UKgal⁻¹ to kg l⁻¹]   1 lb UKgal⁻¹ = 0.099 776 331 kg l⁻¹

(a)

lb UKgal⁻¹	0	10	20	30	40	50	60	70	80	90
0	—	0.997763	1.995527	2.993290	3.991053	4.988817	5.986580	6.984343	7.982106	8.979870

(b)

lb UKgal⁻¹	0	0.1	0.2	0.3	0.4	0.5	0.6	0.7	0.8	0.9
0	—	0.009978	0.019955	0.029933	0.039911	0.049888	0.059866	0.069843	0.079821	0.089799
1	0.099776	0.109754	0.119732	0.129709	0.139687	0.149664	0.159642	0.169620	0.179597	0.189575
2	0.199553	0.209530	0.219508	0.229486	0.239463	0.249441	0.259418	0.269396	0.279374	0.289351
3	0.299329	0.309307	0.319284	0.329262	0.339240	0.349217	0.359195	0.369172	0.379150	0.389128
4	0.399105	0.409083	0.419061	0.429038	0.439016	0.448993	0.458971	0.468949	0.478926	0.488904
5	0.498882	0.508859	0.518837	0.528815	0.538792	0.548770	0.558747	0.568725	0.578703	0.588680
6	0.598658	0.608636	0.618613	0.628591	0.638569	0.648546	0.658524	0.668501	0.678479	0.688457
7	0.698434	0.708412	0.718390	0.728367	0.738345	0.748322	0.758300	0.768278	0.778255	0.788233
8	0.798211	0.808188	0.818166	0.828144	0.838121	0.848099	0.858076	0.868054	0.878032	0.888009
9	0.897987	0.907965	0.917942	0.927920	0.937898	0.947875	0.957853	0.967830	0.977808	0.987786

Example: 17.2 lb UKgal⁻¹ = 0.997763 + 0.718390 kg l⁻¹ = 1.716153 kg l⁻¹

Table 80 KILOGRAMS PER LITRE TO POUNDS PER UK GALLON
[kg l⁻¹ to lb UKgal⁻¹]   1 kg l⁻¹ = 10.022 417 lb UKgal⁻¹

(a)

kg l⁻¹	0	10	20	30	40	50	60	70	80	90
0	—	100.224	200.448	300.673	400.897	501.121	601.345	701.569	801.793	902.018

(b)

kg l⁻¹	0	0.1	0.2	0.3	0.4	0.5	0.6	0.7	0.8	0.9
0	—	1.002	2.004	3.007	4.009	5.011	6.013	7.016	8.018	9.020
1	10.022	11.025	12.027	13.029	14.031	15.034	16.036	17.038	18.040	19.043
2	20.045	21.047	22.049	23.052	24.054	25.056	26.058	27.061	28.063	29.065
3	30.067	31.069	32.072	33.074	34.076	35.078	36.081	37.083	38.085	39.087
4	40.090	41.092	42.094	43.096	44.099	45.101	46.103	47.105	48.108	49.110
5	50.112	51.114	52.117	53.119	54.121	55.123	56.126	57.128	58.130	59.132
6	60.135	61.137	62.139	63.141	64.143	65.146	66.148	67.150	68.152	69.155
7	70.157	71.159	72.161	73.164	74.166	75.168	76.170	77.173	78.175	79.177
8	80.179	81.182	82.184	83.186	84.188	85.191	86.193	87.195	88.197	89.200
9	90.202	91.204	92.206	93.208	94.211	95.213	96.215	97.217	98.220	99.222

Example: 13.3 kg l⁻¹ = 100.224 + 33.074 lb UKgal⁻¹ = 133.298 lb UKgal⁻¹

Table 81 POUNDS-FORCE TO NEWTONS
[lbf to N]   1 lbf = 4.448 221 615 N

(a)

pounds force	0	100	200	300	400	500	600	700	800	900
0	—	444.822	889.644	1334.466	1779.289	2224.111	2668.933	3113.755	3558.577	4003.399

(b)

pounds force	0	1	2	3	4	5	6	7	8	9
0	—	4.448	8.896	13.345	17.793	22.241	26.689	31.138	35.586	40.034
10	44.482	48.930	53.379	57.827	62.275	66.723	71.172	75.620	80.068	84.516
20	88.964	93.413	97.861	102.309	106.757	111.206	115.654	120.102	124.550	128.998
30	133.447	137.895	142.343	146.791	151.240	155.688	160.136	164.584	169.032	173.481
40	177.929	182.377	186.825	191.274	195.722	200.170	204.618	209.066	213.515	217.963
50	222.411	226.859	231.308	235.756	240.204	244.652	249.100	253.549	257.997	262.445
60	266.893	271.342	275.790	280.238	284.686	289.134	293.583	298.031	302.479	306.927
70	311.376	315.824	320.272	324.720	329.168	333.617	338.065	342.513	346.961	351.410
80	355.858	360.306	364.754	369.202	373.651	378.099	382.547	386.995	391.444	395.892
90	400.340	404.788	409.236	413.685	418.133	422.581	427.029	431.477	435.926	440.374

Example: 403 lbf = 1779.289 + 13.345 N = 1792.634 N

Table 82 NEWTONS TO POUNDS-FORCE
[N to lbf]   1 N = 0.224 808 94 lbf

(a)

newtons	0	100	200	300	400	500	600	700	800	900
0	—	22.4809	44.9618	67.4427	89.9236	112.4045	134.8854	157.3663	179.8472	202.3280

(b)

newtons	0	1	2	3	4	5	6	7	8	9
0	—	0.2248	0.4496	0.6744	0.8992	1.1240	1.3489	1.5737	1.7985	2.0233
10	2.2481	2.4729	2.6977	2.9225	3.1473	3.3721	3.5969	3.8218	4.0466	4.2714
20	4.4962	4.7210	4.9458	5.1706	5.3954	5.6202	5.8450	6.0698	6.2947	6.5195
30	6.7443	6.9691	7.1939	7.4187	7.6435	7.8683	8.0931	8.3179	8.5427	8.7675
40	8.9924	9.2172	9.4420	9.6668	9.8916	10.1164	10.3412	10.5660	10.7908	11.0156
50	11.2404	11.4653	11.6901	11.9149	12.1397	12.3645	12.5893	12.8141	13.0389	13.2637
60	13.4885	13.7133	13.9382	14.1630	14.3878	14.6126	14.8374	15.0622	15.2870	15.5118
70	15.7366	15.9614	16.1862	16.4111	16.6359	16.8607	17.0855	17.3103	17.5351	17.7599
80	17.9847	18.2095	18.4343	18.6591	18.8840	19.1088	19.3336	19.5584	19.7832	20.0080
90	20.2328	20.4576	20.6824	20.9072	21.1320	21.3568	21.5817	21.8065	22.0313	22.2561

Example: 812 N = 179.8472 + 2.6977 lbf = 182.5449 lbf
Note: For Tables 81, 82, 83, 84, 85, 86, 87, 88, 97 and 98, see footnote on *Force* on page 6

Table 83 KILOGRAMS-FORCE TO NEWTONS
[kgf to N]   1 kgf = 9.806 65 N (exactly)   [But see footnote on Force on page 6]
All values in Table 69(a) are exact

(a)

kilo-grams force	0	100	200	300	400	500	600	700	800	900
0	—	980.665	1961.330	2941.995	3922.660	4903.325	5883.990	6864.655	7845.320	8825.985

(b)

kilo-grams force	0	1	2	3	4	5	6	7	8	9
0	—	9.807	19.613	29.420	39.227	49.033	58.840	68.647	78.453	88.260
10	98.067	107.873	117.680	127.486	137.293	147.100	156.906	166.713	176.520	186.326
20	196.133	205.940	215.746	225.553	235.360	245.166	254.973	264.780	274.586	284.393
30	294.200	304.006	313.813	323.619	333.426	343.233	353.039	362.846	372.653	382.459
40	392.266	402.073	411.879	421.686	431.493	441.299	451.106	460.913	470.719	480.526
50	490.333	500.139	509.946	519.752	529.559	539.366	549.172	558.979	568.786	578.592
60	588.399	598.206	608.012	617.819	627.626	637.432	647.239	657.046	666.852	676.659
70	686.466	696.272	706.079	715.885	725.692	735.499	745.305	755.112	764.919	774.725
80	784.532	794.339	804.145	813.952	823.759	833.565	843.372	853.179	862.985	872.792
90	882.599	892.405	902.212	912.018	921.825	931.632	941.438	951.245	961.052	970.858

Example: 267 kgf = 1961.330 + 657.046 N = 2618.376 N

Table 84 NEWTONS TO KILOGRAMS-FORCE
[N to kgf]   1 N = 0.101 971 621 kgf

(a)

newtons	0	100	200	300	400	500	600	700	800	900
0	—	10.19716	20.39432	30.59149	40.78865	50.98581	61.18297	71.38013	81.57730	91.77446

(b)

newtons	0	1	2	3	4	5	6	7	8	9
0	—	0.10197	0.20394	0.30591	0.40789	0.50986	0.61183	0.71380	0.81577	0.91774
10	1.01972	1.12169	1.22366	1.32563	1.42760	1.52957	1.63155	1.73352	1.83549	1.93746
20	2.03943	2.14140	2.24338	2.34535	2.44732	2.54929	2.65126	2.75323	2.85521	2.95718
30	3.05915	3.16112	3.26309	3.36506	3.46704	3.56901	3.67098	3.77295	3.87492	3.97689
40	4.07886	4.18084	4.28281	4.38478	4.48675	4.58872	4.69069	4.79267	4.89464	4.99661
50	5.09858	5.20055	5.30252	5.40450	5.50647	5.60844	5.71041	5.81238	5.91435	6.01633
60	6.11830	6.22027	6.32224	6.42421	6.52618	6.62816	6.73013	6.83210	6.93407	7.03604
70	7.13801	7.23999	7.34196	7.44393	7.54590	7.64787	7.74984	7.85181	7.95379	8.05576
80	8.15773	8.25970	8.36167	8.46364	8.56562	8.66759	8.76956	8.87153	8.97350	9.07547
90	9.17745	9.27942	9.38139	9.48336	9.58533	9.68730	9.78928	9.89125	9.99322	10.09519

Example: 753 N = 71.38013 + 5.40450 kgf = 76.78463 kgf
Note: The above tables can also be used to determine Energy (Work, Heat) and Moment of Force (Torque), that is, kilogram-force metres to joules or newton metres (Table 83), and joules or newton metres to kilogram-force metres (Table 84)
Example: 11 kgf m = 107.873 J = 107.873 N m

Table 85 UK TONS-FORCE PER SQUARE INCH TO MEGANEWTONS PER SQUARE METRE

[UK tonf in^{-2} to MN m^{-2}]   I UK tonf in^{-2} = 15.444 256 337 MN m^{-2}

(a)

UK tonf in^{-2}	0	10	20	30	40	50	60	70	80	90
0	—	154.4426	308.8851	463.3277	617.7703	772.2128	926.6554	1081.0979	1235.5405	1389.9831

(b)

UK tonf in^{-2}	0	0.1	0.2	0.3	0.4	0.5	0.6	0.7	0.8	0.9
0	—	1.5444	3.0889	4.6333	6.1777	7.7221	9.2666	10.8110	12.3554	13.8998
1	15.4443	16.9887	18.5331	20.0775	21.6220	23.1664	24.7108	26.2552	27.7997	29.3441
2	30.8885	32.4329	33.9774	35.5218	37.0662	38.6106	40.1551	41.6995	43.2439	44.7883
3	46.3328	47.8772	49.4216	50.9660	52.5105	54.0549	55.5993	57.1437	58.6882	60.2326
4	61.7770	63.3215	64.8659	66.4103	67.9547	69.4992	71.0436	72.5880	74.1324	75.6769
5	77.2213	78.7657	80.3101	81.8546	83.3990	84.9434	86.4878	88.0323	89.5767	91.1211
6	92.6655	94.2100	95.7544	97.2988	98.8432	100.3877	101.9321	103.4765	105.0209	106.5654
7	108.1098	109.6542	111.1986	112.7431	114.2875	115.8319	117.3763	118.9208	120.4652	122.0096
8	123.5541	125.0985	126.6429	128.1873	129.7318	131.2762	132.8206	134.3650	135.9095	137.4539
9	138.9983	140.5427	142.0872	143.6316	145.1760	146.7204	148.2649	149.8093	151.3537	152.8981

Example: 72.6 UK tonf in^{-2} = 1081.0979 + 40.1551 MN m^{-2}
= 1121.2530 MN m^{-2}

Note: The pascal is a special name, recently adopted, for newton per square metre. Thus 1 MN m^{-2} = 1 MPa

Table 86 MEGANEWTONS PER SQUARE METRE TO UK TONS-FORCE PER SQUARE INCH

[MN m^{-2} to UK tonf in^{-2}]   1 MN m^{-2} = 0.064 748 990 1 UK tonf in^{-2}

(a)

MN m^{-2}	0	100	200	300	400	500	600	700	800	900
0	—	6.47490	12.94980	19.42470	25.89960	32.37450	38.84939	45.32429	51.79919	58.27409

(b)

MN m^{-2}	0	1	2	3	4	5	6	7	8	9
0	—	0.06475	0.12950	0.19425	0.25900	0.32374	0.38849	0.45324	0.51799	0.58274
10	0.64749	0.71224	0.77699	0.84174	0.90649	0.97123	1.03598	1.10073	1.16548	1.23023
20	1.29498	1.35973	1.42448	1.48923	1.55398	1.61872	1.68347	1.74822	1.81297	1.87772
30	1.94247	2.00722	2.07197	2.13672	2.20147	2.26621	2.33096	2.39571	2.46046	2.52521
40	2.58996	2.65471	2.71946	2.78421	2.84896	2.91370	2.97845	3.04320	3.10795	3.17270
50	3.23745	3.30220	3.36695	3.43170	3.49645	3.56119	3.62594	3.69069	3.75544	3.82019
60	3.88494	3.94969	4.01444	4.07919	4.14394	4.20868	4.27343	4.33818	4.40293	4.46768
70	4.53243	4.59718	4.66193	4.72668	4.79143	4.85617	4.92092	4.98567	5.05042	5.11517
80	5.17992	5.24467	5.30942	5.37417	5.43892	5.50366	5.56841	5.63316	5.69791	5.76266
90	5.82741	5.89216	5.95691	6.02166	6.08641	6.15115	6.21590	6.28065	6.34540	6.41015

Example: 179 MN m^{-2} = 6.47490 + 5.11517 UK tonf in^{-2} = 11.59007 UK tonf in^{-2}

Note: An alternative metric unit for pressure or stress is the hectobar (hbar) and 1 hbar = 10 MN m^{-2}. Thus: (a) *for the conversion of UK tons-force per square inch to hectobars* divide the values in Table 85 by 10. (b) *For the conversion of hectobars to UK tons-force per square inch* multiply the values in hectobars by 10 thus giving values in meganewtons per square metre and then use Table 86

Table 87 POUNDS-FORCE PER SQUARE INCH TO KILONEWTONS PER SQUARE METRE

[lbf in^{-2} to kN m^{-2}]   1 lbf in^{-2} = 6.894 757 293 kN m^{-2}

(a)

lbf in^{-2}	0	100	200	300	400	500	600	700	800	900
0	—	689.476	1378.951	2068.427	2757.903	3447.379	4136.854	4826.330	5515.806	6205.282

(b)

lbf in^{-2}	0	1	2	3	4	5	6	7	8	9
0	—	6.895	13.790	20.684	27.579	34.474	41.369	48.263	55.158	62.053
10	68.948	75.842	82.737	89.632	96.527	103.421	110.316	117.211	124.106	131.000
20	137.895	144.790	151.685	158.579	165.474	172.369	179.264	186.158	193.053	199.948
30	206.843	213.737	220.632	227.527	234.422	241.317	248.211	255.106	262.001	268.896
40	275.790	282.685	289.580	296.475	303.369	310.264	317.159	324.054	330.948	337.843
50	344.738	351.633	358.527	365.422	372.317	379.212	386.106	393.001	399.896	406.791
60	413.685	420.580	427.475	434.370	441.264	448.159	455.054	461.949	468.843	475.738
70	482.633	489.528	496.423	503.317	510.212	517.107	524.002	530.896	537.791	544.686
80	551.581	558.475	565.370	572.265	579.160	586.054	592.949	599.844	606.739	613.633
90	620.528	627.423	634.318	641.212	648.107	655.002	661.897	668.791	675.686	682.581

Example: 311 lbf in^{-2} = 2068.427 + 75.842 kN m^{-2} = 2144.269 kN m^{-2}

Table 88 KILONEWTONS PER SQUARE METRE TO POUNDS-FORCE PER SQUARE INCH

[kN m^{-2} = 1 lbf in^{-2}]   1 kN m^{-2} = 0.145 037 738 lbf in^{-2}

(a)

kN m^{-2}	0	100	200	300	400	500	600	700	800	900
0	—	14.50377	29.00755	43.51132	58.01510	72.51887	87.02264	101.52642	116.03019	130.53396

(b)

kN m^{-2}	0	1	2	3	4	5	6	7	8	9
0	—	0.14504	0.29008	0.43511	0.58015	0.72519	0.87023	1.01526	1.16030	1.30534
10	1.45038	1.59542	1.74045	1.88549	2.03053	2.17557	2.32060	2.46564	2.61068	2.75572
20	2.90075	3.04579	3.19083	3.33587	3.48091	3.62594	3.77098	3.91602	4.06106	4.20609
30	4.35113	4.49617	4.64121	4.78625	4.93128	5.07632	5.22136	5.36640	5.51143	5.65647
40	5.80151	5.94655	6.09158	6.23662	6.38166	6.52670	6.67174	6.81677	6.96181	7.10685
50	7.25189	7.39692	7.54196	7.68700	7.83204	7.97708	8.12211	8.26715	8.41219	8.55723
60	8.70226	8.84730	8.99234	9.13738	9.28242	9.42745	9.57249	9.71753	9.86257	10.00760
70	10.15264	10.29768	10.44272	10.58775	10.73279	10.87783	11.02287	11.16791	11.31294	11.45798
80	11.60302	11.74806	11.89309	12.03813	12.18317	12.32821	12.47325	12.61828	12.76332	12.90836
90	13.05340	13.19843	13.34347	13.48851	13.63355	13.77859	13.92362	14.06866	14.21370	14.35874

Example: 156 kN m^{-2} = 14.50377 + 8.12211 lbf in^{-2} = 22.62588 lbf in^{-2}

Note: (a) *Conversion of pounds-force per square inch to bars.*
Since 1 lbf in^{-2} = 6.89476 kN m^{-2} = 6.89476 x 10^3 N m^{-2} and 1 bar = 10^5 N m^{-2} then 1 lbf in^{-2} = 6.89476 x 10^{-2} bars.
Thus to convert pounds-force per square inch to bars, multiply values in Table 87 by 10^{-2}
Example: 26 lbf in^{-2} = 179.264 x 10^{-2} bars
  =   1.79264 bars
(b) *Conversion of bars to pounds-force per square inch*
1 bar = 10^2 kN m^{-2} = 10^2 x 0.145 037 738 lbf in^{-2}
Thus, to convert bars to pounds-force per square inch, read the left hand column of Table 88 headed kN m^{-2} as if it were headed 'bars' and multiply the values in Table 88 by 10^2.
Example: 18 bars = 2.61068 x 10^2 lbf in^{-2}
  = 261.068 lbf in^{-2}

Table 89 MILLIMETRES OF MERCURY TO KILONEWTONS PER SQUARE METRE

[mmHg to kN m^{-2}]   1 mmHg = 0.133 322 387 415 kN m^{-2} (exactly)

(a)

mm Hg	0	100	200	300	400	500	600	700	800	900
0	—	13.3322	26.6645	39.9967	53.3290	66.6612	79.9934	93.3257	106.6579	119.9901
1000	133.3224	146.6546	159.9869	173.3191	186.6513	199.9836	213.3158	226.6481	239.9803	253.3125

(b)

mm Hg	0	1	2	3	4	5	6	7	8	9
0	—	0.1333	0.2666	0.4000	0.5333	0.6666	0.7999	0.9333	1.0666	1.1999
10	1.3332	1.4665	1.5999	1.7332	1.8665	1.9998	2.1332	2.2665	2.3998	2.5331
20	2.6664	2.7998	2.9331	3.0664	3.1997	3.3331	3.4664	3.5997	3.7330	3.8663
30	3.9997	4.1330	4.2663	4.3996	4.5330	4.6663	4.7996	4.9329	5.0663	5.1996
40	5.3329	5.4662	5.5995	5.7329	5.8662	5.9995	6.1328	6.2662	6.3995	6.5328
50	6.6661	6.7994	6.9328	7.0661	7.1994	7.3327	7.4661	7.5994	7.7327	7.8660
60	7.9993	8.1327	8.2660	8.3993	8.5326	8.6660	8.7993	8.9326	9.0659	9.1992
70	9.3326	9.4659	9.5992	9.7325	9.8659	9.9992	10.1325	10.2658	10.3991	10.5325
80	10.6658	10.7991	10.9324	11.0658	11.1991	11.3324	11.4657	11.5990	11.7324	11.8657
90	11.9990	12.1323	12.2657	12.3990	12.5323	12.6656	12.7989	12.9323	13.0656	13.1989

Example: 753 mmHg = 93.3257 + 7.0661 kN m^{-2} = 100.3918 kN m^{-2}

Table 90 KILONEWTONS PER SQUARE METRE TO MILLIMETRES OF MERCURY

[kN m^{-2} to mmHg]   1 kN m^{-2} = 7.500 615 758 mmHg

(a)

kN m^{-2}	0	100	200	300	400	500	600	700	800	900
0	—	750.0616	1500.1232	2250.1847	3000.2463	3750.3079	4500.3695	5250.4310	6000.4926	6750.5542

(b)

kN m^{-2}	0	1	2	3	4	5	6	7	8	9
0	—	7.5006	15.0012	22.5018	30.0025	37.5031	45.0037	52.5043	60.0049	67.5055
10	75.0062	82.5068	90.0074	97.5080	105.0086	112.5092	120.0099	127.5105	135.0111	142.5117
20	150.0123	157.5129	165.0135	172.5142	180.0148	187.5154	195.0160	202.5166	210.0172	217.5179
30	225.0185	232.5191	240.0197	247.5203	255.0209	262.5216	270.0222	277.5228	285.0234	292.5240
40	300.0246	307.5252	315.0259	322.5265	330.0271	337.5277	345.0283	352.5289	360.0296	367.5302
50	375.0308	382.5314	390.0320	397.5326	405.0333	412.5339	420.0345	427.5351	435.0357	442.5363
60	450.0369	457.5376	465.0382	472.5388	480.0394	487.5400	495.0406	.502.5413	510.0419	517.5425
70	525.0431	532.5437	540.0443	547.5450	555.0456	562.5462	570.0468	577.5474	585.0480	592.5486
80	600.0493	607.5499	615.0505	622.5511	630.0517	637.5523	645.0530	652.5536	660.0542	667.5548
90	675.0554	682.5560	690.0566	697.5573	705.0579	712.5585	720.0591	727.5597	735.0603	742.5610

Example: 189 kN m^{-2} = 750.0616 + 667.5548 mmHg = 1417.6164 mmHg

Note: In 1971 the pascal (Pa) was internationally recognised as the name for the newton per square metre, hence: 1 kN m^{-2} = 1 kPa

Table 91 INCHES OF MERCURY TO MILLIBARS

[inHg to mbar]   1 inHg = 33.863 9 mbar

inHg	0.00	0.01	0.02	0.03	0.04	0.05	0.06	0.07	0.08	0.09
26.0	880.5	880.8	881.1	881.5	881.8	882.2	882.5	882.8	883.2	883.5
26.1	883.8	884.2	884.5	884.9	885.2	885.5	885.9	886.2	886.6	886.9
26.2	887.2	887.6	887.9	888.3	888.6	888.9	889.3	889.6	889.9	890.3
26.3	890.6	891.0	891.3	891.6	892.0	892.3	892.7	893.0	893.3	893.7
26.4	894.0	894.3	894.7	895.0	895.4	895.7	896.0	896.4	896.7	897.1
26.5	897.4	897.7	898.1	898.4	898.7	899.1	899.4	899.8	900.1	900.4
26.6	900.8	901.1	901.5	901.8	902.1	902.5	902.8	903.2	903.5	903.8
26.7	904.2	904.5	904.8	905.2	905.5	905.9	906.2	906.5	906.9	907.2
26.8	907.6	907.9	908.2	908.6	908.9	909.2	909.6	909.9	910.3	910.6
26.9	910.9	911.3	911.6	912.0	912.3	912.6	913.0	913.3	913.6	914.0
27.0	914.3	914.7	915.0	915.3	915.7	916.0	916.4	916.7	917.0	917.4
27.1	917.7	918.1	918.4	918.7	919.1	919.4	919.7	920.1	920.4	920.8
27.2	921.1	921.4	921.8	922.1	922.5	922.8	923.1	923.5	923.8	924.1
27.3	924.5	924.8	925.2	925.5	925.8	926.2	926.5	926.9	927.2	927.5
27.4	927.9	928.2	928.5	928.9	929.2	929.6	929.9	930.2	930.6	930.9
27.5	931.3	931.6	931.9	932.3	932.6	933.0	933.3	933.6	934.0	934.3
27.6	934.6	935.0	935.3	935.7	936.0	936.3	936.7	937.0	937.4	937.7
27.7	938.0	938.4	938.7	939.0	939.4	939.7	940.1	940.4	940.7	941.1
27.8	941.4	941.8	942.1	942.4	942.8	943.1	943.4	943.8	944.1	944.5
27.9	944.8	945.1	945.5	945.8	946.2	946.5	946.8	947.2	947.5	947.9
28.0	948.2	948.5	948.9	949.2	949.5	949.9	950.2	950.6	950.9	951.2
28.1	951.6	951.9	952.3	952.6	952.9	953.3	953.6	953.9	954.3	954.6
28.2	955.0	955.3	955.6	956.0	956.3	956.7	957.0	957.3	957.7	958.0
28.3	958.3	958.7	959.0	959.4	959.7	960.0	960.4	960.7	961.1	961.4
28.4	961.7	962.1	962.4	962.8	963.1	963.4	963.8	964.1	964.4	964.8
28.5	965.1	965.5	965.8	966.1	966.5	966.8	967.2	967.5	967.8	968.2
28.6	968.5	968.8	969.2	969.5	969.9	970.2	970.5	970.9	971.2	971.6
28.7	971.9	972.2	972.6	972.9	973.2	973.6	973.9	974.3	974.6	974.9
28.8	975.3	975.6	976.0	976.3	976.6	977.0	977.3	977.7	978.0	978.3
28.9	978.7	979.0	979.3	979.7	980.0	980.4	980.7	981.0	981.4	981.7
29.0	982.1	982.4	982.7	983.1	983.4	983.7	984.1	984.4	984.8	985.1
29.1	985.4	985.8	986.1	986.5	986.8	987.1	987.5	987.8	988.1	988.5
29.2	988.8	989.2	989.5	989.8	990.2	990.5	990.9	991.2	991.5	991.9
29.3	992.2	992.6	992.9	993.2	993.6	993.9	994.2	994.6	994.9	995.3
29.4	995.6	995.9	996.3	996.6	997.0	997.3	997.6	998.0	998.3	998.6
29.5	999.0	999.3	999.7	1000.0	1000.3	1000.7	1001.0	1001.4	1001.7	1002.0
29.6	1002.4	1002.7	1003.0	1003.4	1003.7	1004.1	1004.4	1004.7	1005.1	1005.4
29.7	1005.8	1006.1	1006.4	1006.8	1007.1	1007.5	1007.8	1008.1	1008.5	1008.8
29.8	1009.1	1009.5	1009.8	1010.2	1010.5	1010.8	1011.2	1011.5	1011.9	1012.2
29.9	1012.5	1012.9	1013.2	1013.5	1013.9	1014.2	1014.6	1014.9	1015.2	1015.6
30.0	1015.9	1016.3	1016.6	1016.9	1017.3	1017.6	1017.9	1018.3	1018.6	1019.0
30.1	1019.3	1019.6	1020.0	1020.3	1020.7	1021.0	1021.3	1021.7	1022.0	1022.4
30.2	1022.7	1023.0	1023.4	1023.7	1024.0	1024.4	1024.7	1025.1	1025.4	1025.7
30.3	1026.1	1026.4	1026.8	1027.1	1027.4	1027.8	1028.1	1028.4	1028.8	1029.1
30.4	1029.5	1029.8	1030.1	1030.5	1030.8	1031.2	1031.5	1031.8	1032.2	1032.5

inHg	0.00	0.01	0.02	0.03	0.04	0.05	0.06	0.07	0.08	0.09
30.5	1032.8	1033.2	1033.5	1033.9	1034.2	1034.5	1034.9	1035.2	1035.6	1035.9
30.6	1036.2	1036.6	1036.9	1037.3	1037.6	1037.9	1038.3	1038.6	1038.9	1039.3
30.7	1039.6	1040.0	1040.3	1040.6	1041.0	1041.3	1041.7	1042.0	1042.3	1042.7
30.8	1043.0	1043.3	1043.7	1044.0	1044.4	1044.7	1045.0	1045.4	1045.7	1046.1
30.9	1046.4	1046.7	1047.1	1047.4	1047.7	1048.1	1048.4	1048.8	1049.1	1049.4
31.0	1049.8	1050.1	1050.5	1050.8	1051.1	1051.5	1051.8	1052.2	1052.5	1052.8
31.1	1053.2	1053.5	1053.8	1054.2	1054.5	1054.9	1055.2	1055.5	1055.9	1056.2
31.2	1056.6	1056.9	1057.2	1057.6	1057.9	1058.2	1058.6	1058.9	1059.3	1059.6
31.3	1059.9	1060.3	1060.6	1061.0	1061.3	1061.6	1062.0	1062.3	1062.6	1063.0
31.4	1063.3	1063.7	1064.0	1064.3	1064.7	1065.0	1065.4	1065.7	1066.0	1066.4
31.5	1066.7	1067.1	1067.4	1067.7	1068.1	1068.4	1068.7	1069.1	1069.4	1069.8
31.6	1070.1	1070.4	1070.8	1071.1	1071.5	1071.8	1072.1	1072.5	1072.8	1073.1
31.7	1073.5	1073.8	1074.2	1074.5	1074.8	1075.2	1075.5	1075.9	1076.2	1076.5
31.8	1076.9	1077.2	1077.5	1077.9	1078.2	1078.6	1078.9	1079.2	1079.6	1079.9
31.9	1080.3	1080.6	1080.9	1081.3	1081.6	1082.0	1082.3	1082.6	1083.0	1083.3

Example: 29.53 inHg = 1000 mbar

Table 92 MILLIBARS TO INCHES OF MERCURY
[mbar to inHg]   1 mbar = 0.029 530 0 inHg

mbar	0	1	2	3	4	5	6	7	8	9
880	25.99	26.02	26.05	26.07	26.10	26.13	26.16	26.19	26.22	26.25
890	26.28	26.31	26.34	26.37	26.40	26.43	26.46	26.49	26.52	26.55
900	26.58	26.61	26.64	26.67	26.70	26.72	26.75	26.78	26.81	26.84
910	26.87	26.90	26.93	26.96	26.99	27.02	27.05	27.08	27.11	27.14
920	27.17	27.20	27.23	27.26	27.29	27.32	27.34	27.37	27.40	27.43
930	27.46	27.49	27.52	27.55	27.58	27.61	27.64	27.67	27.70	27.73
940	27.76	27.79	27.82	27.85	27.88	27.91	27.94	27.96	27.99	28.02
950	28.05	28.08	28.11	28.14	28.17	28.20	28.23	28.26	28.29	28.32
960	28.35	28.38	28.41	28.44	28.47	28.50	28.53	28.56	28.59	28.61
970	28.64	28.67	28.70	28.73	28.76	28.79	28.82	28.85	28.88	28.91
980	28.94	28.97	29.00	29.03	29.06	29.09	29.12	29.15	29.18	29.21
990	29.23	29.26	29.29	29.32	29.35	29.38	29.41	29.44	29.47	29.50
1000	29.53	29.56	29.59	29.62	29.65	29.68	29.71	29.74	29.77	29.80
1010	29.83	29.85	29.88	29.91	29.94	29.97	30.00	30.03	30.06	30.09
1020	30.12	30.15	30.18	30.21	30.24	30.27	30.30	30.33	30.36	30.39
1030	30.42	30.45	30.47	30.50	30.53	30.56	30.59	30.62	30.65	30.68
1040	30.71	30.74	30.77	30.80	30.83	30.86	30.89	30.92	30.95	30.98
1050	31.01	31.04	31.07	31.10	31.12	31.15	31.18	31.21	31.24	31.27
1060	31.30	31.33	31.36	31.39	31.42	31.45	31.48	31.51	31.54	31.57
1070	31.60	31.63	31.66	31.69	31.72	31.74	31.77	31.80	31.83	31.86

Example: 1012 mbar = 29.88 inHg

Note: (i) for inches of mercury to millimetres of mercury use TABLE 1.
(ii) for millimetres of mercury to inches of mercury use TABLE 2.

Table 93 THERMS TO KILOWATT HOURS

[therms to kW h]   1 therm = 29.307 107 02 kW h

(a)

therms	0	100	200	300	400	500	600	700	800	900
0	—	2930.71	5861.42	8792.13	11722.84	14653.55	17584.26	20514.97	23445.69	26376.40

(b)

therms	0	1	2	3	4	5	6	7	8	9
0	—	29.31	58.61	87.92	117.23	146.54	175.84	205.15	234.46	263.76
10	293.07	322.38	351.69	380.99	410.30	439.61	468.91	498.22	527.53	556.84
20	586.14	615.45	644.76	674.06	703.37	732.68	761.98	791.29	820.60	849.91
30	879.21	908.52	937.83	967.13	996.44	1025.75	1055.06	1084.36	1113.67	1142.98
40	1172.28	1201.59	1230.90	1260.21	1289.51	1318.82	1348.13	1377.43	1406.74	1436.05
50	1465.36	1494.66	1523.97	1553.28	1582.58	1611.89	1641.20	1670.51	1699.81	1729.12
60	1758.43	1787.73	1817.04	1846.35	1875.65	1904.96	1934.27	1963.58	1992.88	2022.19
70	2051.50	2080.80	2110.11	2139.42	2168.73	2198.03	2227.34	2256.65	2285.95	2315.26
80	2344.57	2373.88	2403.18	2432.49	2461.80	2491.10	2520.41	2549.72	2579.03	2608.33
90	2637.64	2666.95	2696.25	2725.56	2754.87	2784.18	2813.48	2842.79	2872.10	2901.40

Example: 789 therms = 20514.97 + 2608.33 kW h = 23123.30 kW h

Table 94 KILOWATT HOURS TO THERMS

[kW h to therms]   1 kW h = 0.034 121 416 therm

(a)

kW h	0	100	200	300	400	500	600	700	800	900
0	—	3.41214	6.82428	10.23642	13.64857	17.06071	20.47285	23.88499	27.29713	30.70927

(b)

kW h	0	1	2	3	4	5	6	7	8	9
0	—	0.03412	0.06824	0.10236	0.13649	0.17061	0.20473	0.23885	0.27297	0.30709
10	0.34121	0.37534	0.40946	0.44358	0.47770	0.51182	0.54594	0.58006	0.61419	0.64831
20	0.68243	0.71655	0.75067	0.78479	0.81891	0.85304	0.88716	0.92128	0.95540	0.98952
30	1.02364	1.05776	1.09189	1.12601	1.16013	1.19425	1.22837	1.26249	1.29661	1.33074
40	1.36486	1.39898	1.43310	1.46722	1.50134	1.53546	1.56959	1.60371	1.63783	1.67195
50	1.70607	1.74019	1.77431	1.80844	1.84256	1.87668	1.91080	1.94492	1.97904	2.01316
60	2.04728	2.08141	2.11553	2.14965	2.18377	2.21789	2.25201	2.28613	2.32026	2.35438
70	2.38850	2.42262	2.45674	2.49086	2.52498	2.55911	2.59323	2.62735	2.66147	2.69559
80	2.72971	2.76383	2.79796	2.83208	2.86620	2.90032	2.93444	2.96856	3.00268	3.03681
90	3.07093	3.10505	3.13917	3.17329	3.20741	3.24153	3.27566	3.30978	3.34390	3.37802

Example: 345 kW h = 10.23642 + 1.53546 therms = 11.77188 therms

Table 95 KILOWATT HOURS TO MEGAJOULES
[kW h to MJ]   1 kW h = 3.6 MJ (exactly)
All values in Table 95 are exact

(a)

kW h	0	100	200	300	400	500	600	700	800	900
0	—	360.000	720.000	1080.000	1440.000	1800.000	2160.000	2520.000	2880.000	3240.000

(b)

kW h	0	1	2	3	4	5	6	7	8	9
0	—	3.600	7.200	10.800	14.400	18.000	21.600	25.200	28.800	32.400
10	36.000	39.600	43.200	46.800	50.400	54.000	57.600	61.200	64.800	68.400
20	72.000	75.600	79.200	82.800	86.400	90.000	93.600	97.200	100.800	104.400
30	108.000	111.600	115.200	118.800	122.400	126.000	129.600	133.200	136.800	140.400
40	144.000	147.600	151.200	154.800	158.400	162.000	165.600	169.200	172.800	176.400
50	180.000	183.600	187.200	190.800	194.400	198.000	201.600	205.200	208.800	212.400
60	216.000	219.600	223.200	226.800	230.400	234.000	237.600	241.200	244.800	248.400
70	252.000	255.600	259.200	262.800	266.400	270.000	273.600	277.200	280.800	284.400
80	288.000	291.600	295.200	298.800	302.400	306.000	309.600	313.200	316.800	320.400
90	324.000	327.600	331.200	334.800	338.400	342.000	345.600	349.200	352.800	356.400

Example: 742 kW h = 2520.0 + 151.2 MJ = 2671.2 MJ

Table 96 MEGAJOULES TO KILOWATT HOURS
[MJ to kW h]   1 MJ = 0.277 777 778 kW h

(a)

MJ	0	100	200	300	400	500	600	700	800	900
0	—	27.7778	55.5556	83.3333	111.1111	138.8889	166.6667	194.4444	222.2222	250.0000

(b)

MJ	0	1	2	3	4	5	6	7	8	9
0	—	0.2778	0.5556	0.8333	1.1111	1.3889	1.6667	1.9444	2.2222	2.5000
10	2.7778	3.0556	3.3333	3.6111	3.8889	4.1667	4.4444	4.7222	5.0000	5.2778
20	5.5556	5.8333	6.1111	6.3889	6.6667	6.9444	7.2222	7.5000	7.7778	8.0556
30	8.3333	8.6111	8.8889	9.1667	9.4444	9.7222	10.0000	10.2778	10.5556	10.8333
40	11.1111	11.3889	11.6667	11.9444	12.2222	12.5000	12.7778	13.0556	13.3333	13.6111
50	13.8889	14.1667	14.4444	14.7222	15.0000	15.2778	15.5556	15.8333	16.1111	16.3889
60	16.6667	16.9444	17.2222	17.5000	17.7778	18.0556	18.3333	18.6111	18.8889	19.1667
70	19.4444	19.7222	20.0000	20.2778	20.5556	20.8333	21.1111	21.3889	21.6667	21.9444
80	22.2222	22.5000	22.7778	23.0556	23.3333	23.6111	23.8889	24.1667	24.4444	24.7222
90	25.0000	25.2778	25.5556	25.8333	26.1111	26.3889	26.6667	26.9444	27.2222	27.5000

Example: 543 MJ = 138.8889 + 11.9444 kW h = 150.8333 kW h

Table 97 FOOT POUNDS-FORCE TO JOULES

[ft lbf to J]   1 ft lbf = 1.355 817 948 J

(a)

ft lbf	0	100	200	300	400	500	600	700	800	900
0	—	135.5818	271.1636	406.7454	542.3272	677.9090	813.4908	949.0726	1084.6544	1220.2362

(b)

ft lbf	0	1	2	3	4	5	6	7	8	9
0	—	1.3558	2.7116	4.0675	5.4233	6.7791	8.1349	9.4907	10.8465	12.2024
10	13.5582	14.9140	16.2698	17.6256	18.9815	20.3373	21.6931	23.0489	24.4047	25.7605
20	27.1164	28.4722	29.8280	31.1838	32.5396	33.8954	35.2513	36.6071	37.9629	39.3187
30	40.6745	42.0304	43.3862	44.7420	46.0978	47.4536	48.8094	50.1653	51.5211	52.8769
40	54.2327	55.5885	56.9444	58.3002	59.6560	61.0118	62.3676	63.7234	65.0793	66.4351
50	67.7909	69.1467	70.5025	71.8584	73.2142	74.5700	75.9258	77.2816	78.6374	79.9933
60	81.3491	82.7049	84.0607	85.4165	86.7723	88.1282	89.4840	90.8398	92.1956	93.5514
70	94.9073	96.2631	97.6189	98.9747	100.3305	101.6863	103.0422	104.3980	105.7538	107.1096
80	108.4654	109.8213	111.1771	112.5329	113.8887	115.2445	116.6003	117.9562	119.3120	120.6678
90	122.0236	123.3794	124.7353	126.0911	127.4469	128.8027	130.1585	131.5143	132.8702	134.2260

Example: 672 ft lbf = 813.4908 + 97.6189 J = 911.1097 J

Table 98 JOULES TO FOOT POUNDS-FORCE

[J to ft lbf]   1 J = 0.737 562 149 ft lbf

(a)

Joules	0	100	200	300	400	500	600	700	800	900
0	—	73.7562	147.5124	221.2686	295.0249	368.7811	442.5373	516.2935	590.0497	663.8059

(b)

Joules	0	1	2	3	4	5	6	7	8	9
0	—	0.7376	1.4751	2.2127	2.9502	3.6878	4.4254	5.1629	5.9005	6.6381
10	7.3756	8.1132	8.8507	9.5883	10.3259	11.0634	11.8010	12.5386	13.2761	14.0137
20	14.7512	15.4888	16.2264	16.9639	17.7015	18.4391	19.1766	19.9142	20.6517	21.3893
30	22.1269	22.8644	23.6020	24.3396	25.0771	25.8147	26.5522	27.2898	28.0274	28.7649
40	29.5025	30.2400	30.9776	31.7152	32.4527	33.1903	33.9279	34.6654	35.4030	36.1405
50	36.8781	37.6157	38.3532	39.0908	39.8284	40.5659	41.3035	42.0410	42.7786	43.5162
60	44.2537	44.9913	45.7289	46.4664	47.2040	47.9415	48.6791	49.4167	50.1542	50.8918
70	51.6294	52.3669	53.1045	53.8420	54.5796	55.3172	56.0547	56.7923	57.5298	58.2674
80	59.0050	59.7425	60.4801	61.2177	61.9552	62.6928	63.4303	64.1679	64.9055	65.6430
90	66.3806	67.1182	67.8557	68.5933	69.3308	70.0684	70.8060	71.5435	72.2811	73.0187

Example: 943 J = 663.8059 + 31.7152 ft lbf = 695.5211 ft lbf

Table 99 HORSEPOWER TO KILOWATTS
[hp to kW]   1 hp = 0.745 699 872 kW

(a)

horse power	0	100	200	300	400	500	600	700	800	900
0	—	74.5700	149.1400	223.7100	298.2799	372.8499	447.4199	521.9899	596.5599	671.1299

(b)

horse power	0	1	2	3	4	5	6	7	8	9
0	—	0.7457	1.4914	2.2371	2.9828	3.7285	4.4742	5.2199	5.9656	6.7113
10	7.4570	8.2027	8.9484	9.6941	10.4398	11.1855	11.9312	12.6769	13.4226	14.1683
20	14.9140	15.6597	16.4054	17.1511	17.8968	18.6425	19.3882	20.1339	20.8796	21.6253
30	22.3710	23.1167	23.8624	24.6081	25.3538	26.0995	26.8452	27.5909	28.3366	29.0823
40	29.8280	30.5737	31.3194	32.0651	32.8108	33.5565	34.3022	35.0479	35.7936	36.5393
50	37.2850	38.0307	38.7764	39.5221	40.2678	41.0135	41.7592	42.5049	43.2506	43.9963
60	44.7420	45.4877	46.2334	46.9791	47.7248	48.4705	49.2162	49.9619	50.7076	51.4533
70	52.1990	52.9447	53.6904	54.4361	55.1818	55.9275	56.6732	57.4189	58.1646	58.9103
80	59.6560	60.4017	61.1474	61.8931	62.6388	63.3845	64.1302	64.8759	65.6216	66.3673
90	67.1130	67.8587	68.6044	69.3501	70.0958	70.8415	71.5872	72.3329	73.0786	73.8243

Example: 497 hp = 298.2799 + 72.3329 kW = 370.6128 kW

Table 100 KILOWATTS TO HORSEPOWER
[kW to hp]   1 kW = 1.341 022 089 hp

(a)

kilowatts	0	100	200	300	400	500	600	700	800	900
0	—	134.1022	268.2044	402.3066	536.4088	670.5110	804.6133	938.7155	1072.8177	1206.9199

(b)

kilowatts	0	1	2	3	4	5	6	7	8	9
0	—	1.3410	2.6820	4.0231	5.3641	6.7051	8.0461	9.3872	10.7282	12.0692
10	13.4102	14.7512	16.0923	17.4333	18.7743	20.1153	21.4564	22.7974	24.1384	25.4794
20	26.8204	28.1615	29.5025	30.8435	32.1845	33.5256	34.8666	36.2076	37.5486	38.8896
30	40.2307	41.5717	42.9127	44.2537	45.5948	46.9358	48.2768	49.6178	50.9588	52.2999
40	53.6409	54.9819	56.3229	57.6639	59.0050	60.3460	61.6870	63.0280	64.3691	65.7101
50	67.0511	68.3921	69.7331	71.0742	72.4152	73.7562	75.0972	76.4383	77.7793	79.1203
60	80.4613	81.8023	83.1434	84.4844	85.8254	87.1664	88.5075	89.8485	91.1895	92.5305
70	93.8715	95.2126	96.5536	97.8946	99.2356	100.5767	101.9177	103.2587	104.5997	105.9407
80	107.2818	108.6228	109.9638	111.3048	112.6459	113.9869	115.3279	116.6689	118.0099	119.3510
90	120.6920	122.0330	123.3740	124.7151	126.0561	127.3971	128.7381	130.0791	131.4202	132.7612

Example: 892 kW = 1072.8177 + 123.3740 hp = 1196.1917 hp

Note: Table 99 can also be used for converting horsepower hour to kilowatts hour, and Table 100 for converting kilowatts hour to horsepower hour.

Table 101   DEGREES FAHRENHEIT TO DEGREES CELSIUS (CENTIGRADE)

[°F to °C] $°C = \frac{5}{9}(°F - 32)$ or $\frac{(°F - 32)}{1.8}$ (exactly)

High Temperature Table (0°F to 990°F)

°F	0	10	20	30	40	50	60	70	80	90
0	−17.8	−12.2	−6.7	−1.1	+4.4	+10.0	+15.6	+21.1	+26.7	+32.2
100	+37.8	+43.3	+48.9	+54.4	+60.0	+65.6	+71.1	+76.7	+82.2	+87.8
200	93.3	98.9	104.4	110.0	115.6	121.1	126.7	132.2	137.8	143.3
300	148.9	154.4	160.0	165.6	171.1	176.7	182.2	187.8	193.3	198.9
400	204.4	210.0	215.6	221.1	226.7	232.2	237.8	243.3	248.9	254.4
500	+260.0	+265.6	+271.1	+276.7	+282.2	+287.8	+293.3	+298.9	+304.4	+310.0
600	315.6	321.1	326.7	332.2	337.8	343.3	348.9	354.4	360.0	365.6
700	371.1	376.7	382.2	387.8	393.3	398.9	404.4	410.0	415.6	421.1
800	426.7	432.2	437.8	443.3	448.9	454.4	460.0	465.6	471.1	476.7
900	482.2	487.8	493.3	498.9	504.4	510.0	515.6	521.1	526.7	532.2

Low Temperature Table (−19.9°F to 99.9°F)

°F	0.0	0.1	0.2	0.3	0.4	0.5	0.6	0.7	0.8	0.9
−19	−28.3	−28.4	−28.4	−28.5	−28.6	−28.6	−28.7	−28.7	−28.8	−28.8
18	27.8	27.8	27.9	27.9	28.0	28.1	28.1	28.2	28.2	28.3
17	27.2	27.3	27.3	27.4	27.4	27.5	27.6	27.6	27.7	27.7
16	26.7	26.7	26.8	26.8	26.9	26.9	27.0	27.1	27.1	27.2
15	26.1	26.2	26.2	26.3	26.3	26.4	26.4	26.5	26.6	26.6
−14	−25.6	−25.6	−25.7	−25.7	−25.8	−25.8	−25.9	−25.9	−26.0	−26.1
13	25.0	25.1	25.1	25.2	25.2	25.3	25.3	25.4	25.4	25.5
12	24.4	24.5	24.6	24.6	24.7	24.7	24.8	24.8	24.9	24.9
11	23.9	23.9	24.0	24.1	24.1	24.2	24.2	24.3	24.3	24.4
10	23.3	23.4	23.4	23.5	23.6	23.6	23.7	23.7	23.8	23.8
−9	−22.8	−22.8	−22.9	−22.9	−23.0	−23.1	−23.1	−23.2	−23.2	−23.3
8	22.2	22.3	22.3	22.4	22.4	22.5	22.6	22.6	22.7	22.7
7	21.7	21.7	21.8	21.8	21.9	21.9	22.0	22.1	22.1	22.2
6	21.1	21.2	21.2	21.3	21.3	21.4	21.4	21.5	21.6	21.6
5	20.6	20.6	20.7	20.7	20.8	20.8	20.9	20.9	21.0	21.1
−4	−20.0	−20.1	−20.1	−20.2	−20.2	−20.3	−20.3	−20.4	−20.4	−20.5
3	19.4	19.5	19.6	19.6	19.7	19.7	19.8	19.8	19.9	19.9
2	18.9	18.9	19.0	19.1	19.1	19.2	19.2	19.3	19.3	19.4
1	18.3	18.4	18.4	18.5	18.6	18.6	18.7	18.7	18.8	18.8
−0	−17.8	−17.8	−17.9	−17.9	−18.0	−18.1	−18.1	−18.2	−18.2	−18.3
+0	−17.8	−17.7	−17.7	−17.6	−17.6	−17.5	−17.4	−17.4	−17.3	−17.3
1	17.2	17.2	17.1	17.1	17.0	16.9	16.9	16.8	16.8	16.7
2	16.7	16.6	16.6	16.5	16.4	16.4	16.3	16.3	16.2	16.2
3	16.1	16.1	16.0	15.9	15.9	15.8	15.8	15.7	15.7	15.6
4	15.6	15.5	15.4	15.4	15.3	15.3	15.2	15.2	15.1	15.1
+5	−15.0	−14.9	−14.9	−14.8	−14.8	−14.7	−14.7	−14.6	−14.6	−14.5
6	14.4	14.4	14.3	14.3	14.2	14.2	14.1	14.1	14.0	13.9
7	13.9	13.8	13.8	13.7	13.7	13.6	13.6	13.5	13.4	13.4
8	13.3	13.3	13.2	13.2	13.1	13.1	13.0	12.9	12.9	12.8
9	12.8	12.7	12.7	12.6	12.6	12.5	12.4	12.4	12.3	12.3

Note: The *High Temperature Table* is for the conversion of temperatures of household ovens, etc.; the *Low Temperature Table* is for the conversion of climatic statistics, etc. The two tables *must* be used independently. For the conversion of values not shown in these tables, the reader should use the conversion factor given at the head of the tables.

°F	0.0	0.1	0.2	0.3	0.4	0.5	0.6	0.7	0.8	0.9
+10	−12.2	−12.2	−12.1	−12.1	−12.0	−11.9	−11.9	−11.8	−11.8	−11.7
11	11.7	11.6	11.6	11.5	11.4	11.4	11.3	11.3	11.2	11.2
12	11.1	11.1	11.0	10.9	10.9	10.8	10.8	10.7	10.7	10.6
13	10.6	10.5	10.4	10.4	10.3	10.3	10.2	10.2	10.1	10.1
14	10.0	9.9	9.9	9.8	9.8	9.7	9.7	9.6	9.6	9.5
+15	−9.4	−9.4	−9.3	−9.3	−9.2	−9.2	−9.1	−9.1	−9.0	−8.9
16	8.9	8.8	8.8	8.7	8.7	8.6	8.6	8.5	8.4	8.4
17	8.3	8.3	8.2	8.2	8.1	8.1	8.0	7.9	7.9	7.8
18	7.8	7.7	7.7	7.6	7.6	7.5	7.4	7.4	7.3	7.3
19	7.2	7.2	7.1	7.1	7.0	6.9	6.9	6.8	6.8	6.7
+20	−6.7	−6.6	−6.6	−6.5	−6.4	−6.4	−6.3	−6.3	−6.2	−6.2
21	6.1	6.1	6.0	5.9	5.9	5.8	5.8	5.7	5.7	5.6
22	5.6	5.5	5.4	5.4	5.3	5.3	5.2	5.2	5.1	5.1
23	5.0	4.9	4.9	4.8	4.8	4.7	4.7	4.6	4.6	4.5
24	4.4	4.4	4.3	4.3	4.2	4.2	4.1	4.1	4.0	3.9
+25	−3.9	−3.8	−3.8	−3.7	−3.7	−3.6	−3.6	−3.5	−3.4	−3.4
26	3.3	3.3	3.2	3.2	3.1	3.1	3.0	2.9	2.9	2.8
27	2.8	2.7	2.7	2.6	2.6	2.5	2.4	2.4	2.3	2.3
28	2.2	2.2	2.1	2.1	2.0	1.9	1.9	1.8	1.8	1.7
29	1.7	1.6	1.6	1.5	1.4	1.4	1.3	1.3	1.2	1.2
+30	−1.1	−1.1	−1.0	−0.9	−0.9	−0.8	−0.8	−0.7	−0.7	−0.6
31	−0.6	−0.5	−0.4	−0.4	−0.3	−0.3	−0.2	−0.2	−0.1	−0.1
32	0.0	+0.1	+0.1	+0.2	+0.2	+0.3	+0.3	+0.4	+0.4	+0.5
33	+0.6	+0.6	+0.7	+0.7	+0.8	+0.8	+0.9	+0.9	+1.0	+1.1
34	+1.1	+1.2	+1.2	+1.3	+1.3	+1.4	+1.4	+1.5	+1.6	+1.6
+35	+1.7	+1.7	+1.8	+1.8	+1.9	+1.9	+2.0	+2.1	+2.1	+2.2
36	2.2	2.3	2.3	2.4	2.4	2.5	2.6	2.6	2.7	2.7
37	2.8	2.8	2.9	2.9	3.0	3.1	3.1	3.2	3.2	3.3
38	3.3	3.4	3.4	3.5	3.6	3.6	3.7	3.7	3.8	3.8
39	3.9	3.9	4.0	4.1	4.1	4.2	4.2	4.3	4.3	4.4
+40	+4.4	+4.5	+4.6	+4.6	+4.7	+4.7	+4.8	+4.8	+4.9	+4.9
41	5.0	5.1	5.1	5.2	5.2	5.3	5.3	5.4	5.4	5.5
42	5.6	5.6	5.7	5.7	5.8	5.8	5.9	5.9	6.0	6.1
43	6.1	6.2	6.2	6.3	6.3	6.4	6.4	6.5	6.6	6.6
44	6.7	6.7	6.8	6.8	6.9	6.9	7.0	7.1	7.1	7.2
+45	+7.2	+7.3	+7.3	+7.4	+7.4	+7.5	+7.6	+7.6	+7.7	+7.7
46	7.8	7.8	7.9	7.9	8.0	8.1	8.1	8.2	8.2	8.3
47	8.3	8.4	8.4	8.5	8.6	8.6	8.7	8.7	8.8	8.8
48	8.9	8.9	9.0	9.1	9.1	9.2	9.2	9.3	9.3	9.4
49	9.4	9.5	9.6	9.6	9.7	9.7	9.8	0.0	0.0	9.9
+50	+10.0	+10.1	+10.1	+10.2	+10.2	+10.3	+10.3	+10.4	+10.4	+10.5
51	10.6	10.6	10.7	10.7	10.8	10.8	10.9	10.9	11.0	11.1
52	11.1	11.2	11.2	11.3	11.3	11.4	11.4	11.5	11.6	11.6
53	11.7	11.7	11.8	11.8	11.9	11.9	12.0	12.1	12.1	12.2
54	12.2	12.3	12.3	12.4	12.4	12.5	12.6	12.6	12.7	12.7

°F	0.0	0.1	0.2	0.3	0.4	0.5	0.6	0.7	0.8	0.9
+55	+12.8	+12.8	+12.9	+12.9	+13.0	+13.1	+13.1	+13.2	+13.2	+13.3
56	13.3	13.4	13.4	13.5	13.6	13.6	13.7	13.7	13.8	13.8
57	13.9	13.9	14.0	14.1	14.1	14.2	14.2	14.3	14.3	14.4
58	14.4	14.5	14.6	14.6	14.7	14.7	14.8	14.8	14.9	14.9
59	15.0	15.1	15.1	15.2	15.2	15.3	15.3	15.4	15.4	15.5
+60	+15.6	+15.6	+15.7	+15.7	+15.8	+15.8	+15.9	+15.9	+16.0	+16.1
61	16.1	16.2	16.2	16.3	16.3	16.4	16.4	16.5	16.6	16.6
62	16.7	16.7	16.8	16.8	16.9	16.9	17.0	17.1	17.1	17.2
63	17.2	17.3	17.3	17.4	17.4	17.5	17.6	17.6	17.7	17.7
64	17.8	17.8	17.9	17.9	18.0	18.1	18.1	18.2	18.2	18.3
+65	+18.3	+18.4	+18.4	+18.5	+18.6	+18.6	+18.7	+18.7	+18.8	+18.8
66	18.9	18.9	19.0	19.1	19.1	19.2	19.2	19.3	19.3	19.4
67	19.4	19.5	19.6	19.6	19.7	19.7	19.8	19.8	19.9	19.9
68	20.0	20.1	20.1	20.2	20.2	20.3	20.3	20.4	20.4	20.5
69	20.6	20.6	20.7	20.7	20.8	20.8	20.9	20.9	21.0	21.1
+70	+21.1	+21.2	+21.2	+21.3	+21.3	+21.4	+21.4	+21.5	+21.6	+21.6
71	21.7	21.7	21.8	21.8	21.9	21.9	22.0	22.1	22.1	22.2
72	22.2	22.3	22.3	22.4	22.4	22.5	22.6	22.6	22.7	22.7
73	22.8	22.8	22.9	22.9	23.0	23.1	23.1	23.2	23.2	23.3
74	23.3	23.4	23.4	23.5	23.6	23.6	23.7	23.7	23.8	23.8
+75	+23.9	+23.9	+24.0	+24.1	+24.1	+24.2	+24.2	+24.3	+24.3	+24.4
76	24.4	24.5	24.6	24.6	24.7	24.7	24.8	24.8	24.9	24.9
77	25.0	25.1	25.1	25.2	25.2	25.3	25.3	25.4	25.4	25.5
78	25.6	25.6	25.7	25.7	25.8	25.8	25.9	25.9	26.0	26.1
79	26.1	26.2	26.2	26.3	26.3	26.4	26.4	26.5	26.6	26.6
+80	+26.7	+26.7	+26.8	+26.8	+26.9	+26.9	+27.0	+27.1	+27.1	+27.2
81	27.2	27.3	27.3	27.4	27.4	27.5	27.6	27.6	27.7	27.7
82	27.8	27.8	27.9	27.9	28.0	28.1	28.1	28.2	28.2	28.3
83	28.3	28.4	28.4	28.5	28.6	28.6	28.7	28.7	28.8	28.8
84	28.9	28.9	29.0	29.1	29.1	29.2	29.2	29.3	29.3	29.4
+85	+29.4	+29.5	+29.6	+29.6	+29.7	+29.7	+29.8	+29.8	+29.9	+29.9
86	30.0	30.1	30.1	30.2	30.2	30.3	30.3	30.4	30.4	30.5
87	30.6	30.6	30.7	30.7	30.8	30.8	30.9	30.9	31.0	31.1
88	31.1	31.2	31.2	31.3	31.3	31.4	31.4	31.5	31.6	31.6
89	31.7	31.7	31.8	31.8	31.9	31.9	32.0	32.1	32.1	32.2
+90	+32.2	+32.3	+32.3	+32.4	+32.4	+32.5	+32.6	+32.6	+32.7	+32.7
91	32.8	32.8	32.9	32.9	33.0	33.1	33.1	33.2	33.2	33.3
92	33.3	33.4	33.4	33.5	33.6	33.6	33.7	33.7	33.8	33.8
93	33.9	33.9	34.0	34.1	34.1	34.2	34.2	34.3	34.3	34.4
94	34.4	34.5	34.6	34.6	34.7	34.7	34.8	34.8	34.9	34.9
+95	+35.0	+35.1	+35.1	+35.2	+35.2	+35.3	+35.3	+35.4	+35.4	+35.5
96	35.6	35.6	35.7	35.7	35.8	35.8	35.9	35.9	36.0	36.1
97	36.1	36.2	36.2	36.3	36.3	36.4	36.4	36.5	36.6	36.6
98	36.7	36.7	36.8	36.8	36.9	36.9	37.0	37.1	37.1	37.2
99	37.2	37.3	37.3	37.4	37.4	37.5	37.6	37.6	37.7	37.7

Example: 62.5°F = 16.9°C
Note: Conversion of degrees Fahrenheit to Kelvins.
To convert degrees Fahrenheit to Kelvins first convert degrees Fahrenheit to degrees Celsius (Centigrade) and then add 273.15
Example: 77°F = 25°C = 25 + 273.15 K = 298.15 K

Table 102 DEGREES CELSIUS (CENTIGRADE) TO DEGREES FAHRENHEIT

[°C to °F] $°F = \frac{9}{5} °C + 32$ or $(°C \times 1.8) + 32$ (exactly)

High Temperature Table (0°C to 990°C)   All values in this table are exact

°C	0	10	20	30	40	50	60	70	80	90
0	+32.0	+50.0	+68.0	+86.0	+104.0	+122.0	+140.0	+158.0	+176.0	+194.0
100	212.0	230.0	248.0	266.0	284.0	302.0	320.0	338.0	356.0	374.0
200	392.0	410.0	428.0	446.0	464.0	482.0	500.0	518.0	536.0	554.0
300	572.0	590.0	608.0	626.0	644.0	662.0	680.0	698.0	716.0	734.0
400	752.0	770.0	788.0	806.0	824.0	842.0	860.0	878.0	896.0	914.0
500	+932.0	+950.0	+968.0	+986.0	+1004.0	+1022.0	+1040.0	+1058.0	+1076.0	+1094.0
600	1112.0	1130.0	1148.0	1166.0	1184.0	1202.0	1220.0	1238.0	1256.0	1274.0
700	1292.0	1310.0	1328.0	1346.0	1364.0	1382.0	1400.0	1418.0	1436.0	1454.0
800	1472.0	1490.0	1508.0	1526.0	1544.0	1562.0	1580.0	1598.0	1616.0	1634.0
900	1652.0	1670.0	1688.0	1706.0	1724.0	1742.0	1760.0	1778.0	1796.0	1814.0

Low Temperature Table (−34.9°C to 39.9°C)

°C	0.0	0.1	0.2	0.3	0.4	0.5	0.6	0.7	0.8	0.9
−34	−29.2	−29.4	−29.6	−29.7	−29.9	−30.1	−30.3	−30.5	−30.6	−30.8
33	27.4	27.6	27.8	27.9	28.1	28.3	28.5	28.7	28.8	29.0
32	25.6	25.8	26.0	26.1	26.3	26.5	26.7	26.9	27.0	27.2
31	23.8	24.0	24.2	24.3	24.5	24.7	24.9	25.1	25.2	25.4
30	22.0	22.2	22.4	22.5	22.7	22.9	23.1	23.3	23.4	23.6
−29	−20.2	−20.4	−20.6	−20.7	−20.9	−21.1	−21.3	−21.5	−21.6	−21.8
28	18.4	18.6	18.8	18.9	19.1	19.3	19.5	19.7	19.8	20.0
27	16.6	16.8	17.0	17.1	17.3	17.5	17.7	17.9	18.0	18.2
26	14.8	15.0	15.2	15.3	15.5	15.7	15.9	16.1	16.2	16.4
25	13.0	13.2	13.4	13.5	13.7	13.9	14.1	14.3	14.4	14.6
−24	−11.2	−11.4	−11.6	−11.7	−11.9	−12.1	−12.3	−12.5	−12.6	−12.8
23	9.4	9.6	9.8	9.9	10.1	10.3	10.5	10.7	10.8	11.0
22	7.6	7.8	8.0	8.1	8.3	8.5	8.7	8.9	9.0	9.2
21	5.8	6.0	6.2	6.3	6.5	6.7	6.9	7.1	7.2	7.4
20	4.0	4.2	4.4	4.5	4.7	4.9	5.1	5.3	5.4	5.6
−19	−2.2	−2.4	−2.6	−2.7	−2.9	−3.1	−3.3	−3.5	−3.6	−3.8
18	−0.4	−0.6	−0.8	−0.9	−1.1	−1.3	−1.5	−1.7	−1.8	−2.0
17	+1.4	+1.2	+1.0	+0.9	+0.7	+0.5	+0.3	+0.1	0.0	−0.2
16	+3.2	+3.0	+2.8	+2.7	+2.5	+2.3	+2.1	+1.9	+1.8	+1.6
15	5.0	4.8	4.6	4.5	4.3	4.1	3.9	3.7	3.6	3.4
−14	+6.8	+6.6	+6.4	+6.3	+6.1	+5.9	+5.7	+5.5	+5.4	+5.2
13	8.6	8.4	8.2	8.1	7.9	7.7	7.5	7.3	7.2	7.0
12	10.4	10.2	10.0	9.9	9.7	9.5	9.3	9.1	9.0	8.8
11	12.2	12.0	11.8	11.7	11.5	11.3	11.1	10.9	10.8	10.6
10	14.0	13.8	13.6	13.5	13.3	13.1	12.9	12.7	12.6	12.4
−9	+15.8	+15.6	+15.4	+15.3	+15.1	+14.9	+14.7	+14.5	+14.4	+14.2
8	17.6	17.4	17.2	17.1	16.9	16.7	16.5	16.3	16.2	16.0
7	19.4	19.2	19.0	18.9	18.7	18.5	18.3	18.1	18.0	17.8
6	21.2	21.0	20.8	20.7	20.5	20.3	20.1	19.9	19.8	19.6
5	23.0	22.8	22.6	22.5	22.3	22.1	21.9	21.7	21.6	21.4

°C	0.0	0.1	0.2	0.3	0.4	0.5	0.6	0.7	0.8	0.9
−4	+24.8	+24.6	+24.4	+24.3	+24.1	+23.9	+23.7	+23.5	+23.4	+23.2
3	26.6	26.4	26.2	26.1	25.9	25.7	25.5	25.3	25.2	25.0
2	28.4	28.2	28.0	27.9	27.7	27.5	27.3	27.1	27.0	26.8
1	30.2	30.0	29.8	29.7	29.5	29.3	29.1	28.9	28.8	28.6
−0	+32.0	+31.8	+31.6	+31.5	+31.3	+31.1	+30.9	+30.7	+30.6	+30.4
+0	+32.0	+32.2	+32.4	+32.5	+32.7	+32.9	+33.1	+33.3	+33.4	+33.6
1	33.8	34.0	34.2	34.3	34.5	34.7	34.9	35.1	35.2	35.4
2	35.6	35.8	36.0	36.1	36.3	36.5	36.7	36.9	37.0	37.2
3	37.4	37.6	37.8	37.9	38.1	38.3	38.5	38.7	38.8	39.0
4	39.2	39.4	39.6	39.7	39.9	40.1	40.3	40.5	40.6	40.8
+5	+41.0	+41.2	+41.4	+41.5	+41.7	+41.9	+42.1	+42.3	+42.4	+42.6
6	42.8	43.0	43.2	43.3	43.5	43.7	43.9	44.1	44.2	44.4
7	44.6	44.8	45.0	45.1	45.3	45.5	45.7	45.9	46.0	46.2
8	46.4	46.6	46.8	46.9	47.1	47.3	47.5	47.7	47.8	48.0
9	48.2	48.4	48.6	48.7	48.9	49.1	49.3	49.5	49.6	49.8
+10	+50.0	+50.2	+50.4	+50.5	+50.7	+50.9	+51.1	+51.3	+51.4	+51.6
11	51.8	52.0	52.2	52.3	52.5	52.7	52.9	53.1	53.2	53.4
12	53.6	53.8	54.0	54.1	54.3	54.5	54.7	54.9	55.0	55.2
13	55.4	55.6	55.8	55.9	56.1	56.3	56.5	56.7	56.8	57.0
14	57.2	57.4	57.6	57.7	57.9	58.1	58.3	58.5	58.6	58.8
+15	+59.0	+59.2	+59.4	+59.5	+59.7	+59.9	+60.1	+60.3	+60.4	+60.6
16	60.8	61.0	61.2	61.3	61.5	61.7	61.9	62.1	62.2	62.4
17	62.6	62.8	63.0	63.1	63.3	63.5	63.7	63.9	64.0	64.2
18	64.4	64.6	64.8	64.9	65.1	65.3	65.5	65.7	65.8	66.0
19	66.2	66.4	66.6	66.7	66.9	67.1	67.3	67.5	67.6	67.8
+20	+68.0	+68.2	+68.4	+68.5	+68.7	+68.9	+69.1	+69.3	+69.4	+69.6
21	69.8	70.0	70.2	70.3	70.5	70.7	70.9	71.1	71.2	71.4
22	71.6	71.8	72.0	72.1	72.3	72.5	72.7	72.9	73.0	73.2
23	73.4	73.6	73.8	73.9	74.1	74.3	74.5	74.7	74.8	75.0
24	75.2	75.4	75.6	75.7	75.9	76.1	76.3	76.5	76.6	76.8
+25	+77.0	+77.2	+77.4	+77.5	+77.7	+77.9	+78.1	+78.3	+78.4	+78.6
26	78.8	79.0	79.2	79.3	79.5	79.7	79.9	80.1	80.2	80.4
27	80.6	80.8	81.0	81.1	81.3	81.5	81.7	81.9	82.0	82.2
28	82.4	82.6	82.8	82.9	83.1	83.3	83.5	83.7	83.8	84.0
29	84.2	84.4	84.6	84.7	84.9	85.1	85.3	85.5	85.6	85.8
+30	+86.0	+86.2	+86.4	+86.5	+86.7	+86.9	+87.1	+87.3	+87.4	+87.6
31	87.8	88.0	88.2	88.3	88.5	88.7	88.9	89.1	89.2	89.4
32	89.6	89.8	90.0	90.1	90.3	90.5	90.7	90.9	91.0	91.2
33	91.4	91.6	91.8	91.9	92.1	92.3	92.5	92.7	92.8	93.0
34	93.2	93.4	93.6	93.7	93.9	94.1	94.3	94.5	94.6	94.8
+35	+95.0	+95.2	+95.4	+95.5	+95.7	+95.9	+96.1	+96.3	+96.4	+96.6
36	96.8	97.0	97.2	97.3	97.5	97.7	97.9	98.1	98.2	98.4
37	98.6	98.8	99.0	99.1	99.3	99.5	99.7	99.9	100.0	100.2
38	100.4	100.6	100.8	100.9	101.1	101.3	101.5	101.7	101.8	102.0
39	102.2	102.4	102.6	102.7	102.9	103.1	103.3	103.5	103.6	103.8

Example: 4.3°C = 39.7°F
Note: Conversion of degrees Celsius (Centigrade) to Kelvins
To convert degrees Celsius (Centigrade) to Kelvins add 273.15
Example: 30°C = 30 + 273.15 K
= 303.15 K

Table 103 DIFFERENCES: DEGREES FAHRENHEIT TO DEGREES CELSIUS (CENTIGRADE)

[°F to °C] 1°F = 0.555 6°C

°F	0.0	0.1	0.2	0.3	0.4	0.5	0.6	0.7	0.8	0.9
0	—	0.1	0.1	0.2	0.2	0.3	0.3	0.4	0.4	0.5
1	0.6	0.6	0.7	0.7	0.8	0.8	0.9	0.9	1.0	1.1
2	1.1	1.2	1.2	1.3	1.3	1.4	1.4	1.5	1.6	1.6
3	1.7	1.7	1.8	1.8	1.9	1.9	2.0	2.1	2.1	2.2
4	2.2	2.3	2.3	2.4	2.4	2.5	2.6	2.6	2.7	2.7
5	2.8	2.8	2.9	2.9	3.0	3.1	3.1	3.2	3.2	3.3
6	3.3	3.4	3.4	3.5	3.6	3.6	3.7	3.7	3.8	3.8
7	3.9	3.9	4.0	4.1	4.1	4.2	4.2	4.3	4.3	4.4
8	4.4	4.5	4.6	4.6	4.7	4.7	4.8	4.8	4.9	4.9
9	5.0	5.1	5.1	5.2	5.2	5.3	5.3	5.4	5.4	5.5
10	5.6	5.6	5.7	5.7	5.8	5.8	5.9	5.9	6.0	6.1
11	6.1	6.2	6.2	6.3	6.3	6.4	6.4	6.5	6.6	6.6
12	6.7	6.7	6.8	6.8	6.9	6.9	7.0	7.1	7.1	7.2
13	7.2	7.3	7.3	7.4	7.4	7.5	7.6	7.6	7.7	7.7
14	7.8	7.8	7.9	7.9	8.0	8.1	8.1	8.2	8.2	8.3
15	8.3	8.4	8.4	8.5	8.6	8.6	8.7	8.7	8.8	8.8
16	8.9	8.9	9.0	9.1	9.1	9.2	9.2	9.3	9.3	9.4
17	9.4	9.5	9.6	9.6	9.7	9.7	9.8	9.8	9.9	9.9
18	10.0	10.1	10.1	10.2	10.2	10.3	10.3	10.4	10.4	10.5
19	10.6	10.6	10.7	10.7	10.8	10.8	10.9	10.9	11.0	11.1
20	11.1	11.2	11.2	11.3	11.3	11.4	11.4	11.5	11.6	11.6
21	11.7	11.7	11.8	11.8	11.9	11.9	12.0	12.1	12.1	12.2
22	12.2	12.3	12.3	12.4	12.4	12.5	12.6	12.6	12.7	12.7
23	12.8	12.8	12.9	12.9	13.0	13.1	13.1	13.2	13.2	13.3
24	13.3	13.4	13.4	13.5	13.6	13.6	13.7	13.7	13.8	13.8
25	13.9	13.9	14.0	14.1	14.1	14.2	14.2	14.3	14.3	14.4
26	14.4	14.5	14.6	14.6	14.7	14.7	14.8	14.8	14.9	14.9
27	15.0	15.1	15.1	15.2	15.2	15.3	15.3	15.4	15.4	15.5
28	15.6	15.6	15.7	15.7	15.8	15.8	15.9	15.9	16.0	16.1
29	16.1	16.2	16.2	16.3	16.3	16.4	16.4	16.5	16.6	16.6
30	16.7	16.7	16.8	16.8	16.9	16.9	17.0	17.1	17.1	17.2
31	17.2	17.3	17.3	17.4	17.4	17.5	17.6	17.6	17.7	17.7
32	17.8	17.8	17.9	17.9	18.0	18.1	18.1	18.2	18.2	18.3
33	18.3	18.4	18.4	18.5	18.6	18.6	18.7	18.7	18.8	18.8
34	18.9	18.9	19.0	19.1	19.1	19.2	19.2	19.3	19.3	19.4
35	19.4	19.5	19.6	19.6	19.7	19.7	19.8	19.8	19.9	19.9
36	20.0	20.1	20.1	20.2	20.2	20.3	20.3	20.4	20.4	20.5
37	20.6	20.6	20.7	20.7	20.8	20.8	20.9	20.9	21.0	21.1
38	21.1	21.2	21.2	21.3	21.3	21.4	21.4	21.5	21.6	21.6
39	21.7	21.7	21.8	21.8	21.9	21.9	22.0	22.1	22.1	22.2
40	22.2	22.3	22.3	22.4	22.4	22.5	22.6	22.6	22.7	22.7
41	22.8	22.8	22.9	22.9	23.0	23.1	23.1	23.2	23.2	23.3
42	23.3	23.4	23.4	23.5	23.6	23.6	23.7	23.7	23.8	23.8
43	23.9	23.9	24.0	24.1	24.1	24.2	24.2	24.3	24.3	24.4
44	24.4	24.5	24.6	24.6	24.7	24.7	24.8	24.8	24.9	24.9

°F to °C

°F	0.0	0.1	0.2	0.3	0.4	0.5	0.6	0.7	0.8	0.9
45	25.0	25.1	25.1	25.2	25.2	25.3	25.3	25.4	25.4	25.5
46	25.6	25.6	25.7	25.7	25.8	25.8	25.9	25.9	26.0	26.1
47	26.1	26.2	26.2	26.3	26.3	26.4	26.4	26.5	26.6	26.6
48	26.7	26.7	26.8	26.8	26.9	26.9	27.0	27.1	27.1	27.2
49	27.2	27.3	27.3	27.4	27.4	27.5	27.6	27.6	27.7	27.7
50	27.8	27.8	27.9	27.9	28.0	28.1	28.1	28.2	28.2	28.3
51	28.3	28.4	28.4	28.5	28.6	28.6	28.7	28.7	28.8	28.8
52	28.9	28.9	29.0	29.1	29.1	29.2	29.2	29.3	29.3	29.4
53	29.4	29.5	29.6	29.6	29.7	29.7	29.8	29.8	29.9	29.9
54	30.0	30.1	30.1	30.2	30.2	30.3	30.3	30.4	30.4	30.5
55	30.6	30.6	30.7	30.7	30.8	30.8	30.9	30.9	31.0	31.1
56	31.1	31.2	31.2	31.3	31.3	31.4	31.4	31.5	31.6	31.6
57	31.7	31.7	31.8	31.8	31.9	31.9	32.0	32.1	32.1	32.2
58	32.2	32.3	32.3	32.4	32.4	32.5	32.6	32.6	32.7	32.7
59	32.8	32.8	32.9	32.9	33.0	33.1	33.1	33.2	33.2	33.3
60	33.3	33.4	33.4	33.5	33.6	33.6	33.7	33.7	33.8	33.8
61	33.9	33.9	34.0	34.1	34.1	34.2	34.2	34.3	34.3	34.4
62	34.4	34.5	34.6	34.6	34.7	34.7	34.8	34.8	34.9	34.9
63	35.0	35.1	35.1	35.2	35.2	35.3	35.3	35.4	35.4	35.5
64	35.6	35.6	35.7	35.7	35.8	35.8	35.9	35.9	36.0	36.1
65	36.1	36.2	36.2	36.3	36.3	36.4	36.4	36.5	36.6	36.6
66	36.7	36.7	36.8	36.8	36.9	36.9	37.0	37.1	37.1	37.2
67	37.2	37.3	37.3	37.4	37.4	37.5	37.6	37.6	37.7	37.7
68	37.8	37.8	37.9	37.9	38.0	38.1	38.1	38.2	38.2	38.3
69	38.3	38.4	38.4	38.5	38.6	38.6	38.7	38.7	38.8	38.8
70	38.9	38.9	39.0	39.1	39.1	39.2	39.2	39.3	39.3	39.4
71	39.4	39.5	39.6	39.6	39.7	39.7	39.8	39.8	39.9	39.9
72	40.0	40.1	40.1	40.2	40.2	40.3	40.3	40.4	40.4	40.5
73	40.6	40.6	40.7	40.7	40.8	40.8	40.9	40.9	41.0	41.1
74	41.1	41.2	41.2	41.3	41.3	41.4	41.4	41.5	41.6	41.6
75	41.7	41.7	41.8	41.8	41.9	41.9	42.0	42.1	42.1	42.2
76	42.2	42.3	42.3	42.4	42.4	42.5	42.6	42.6	42.7	42.7
77	42.8	42.8	42.9	42.9	43.0	43.1	43.1	43.2	43.2	43.3
78	43.3	43.4	43.4	43.5	43.6	43.6	43.7	43.7	43.8	43.8
79	43.9	43.9	44.0	44.1	44.1	44.2	44.2	44.3	44.3	44.4
80	44.4	44.5	44.6	44.6	44.7	44.7	44.8	44.8	44.9	44.9
81	45.0	45.1	45.1	45.2	45.2	45.3	45.3	45.4	45.4	45.5
82	45.6	45.6	45.7	45.7	45.8	45.8	45.9	45.9	46.0	46.1
83	46.1	46.2	46.2	46.3	46.3	46.4	46.4	46.5	46.6	46.6
84	46.7	46.7	46.8	46.8	46.9	46.9	47.0	47.1	47.1	47.2
85	47.2	47.3	47.3	47.4	47.4	47.5	47.6	47.6	47.7	47.7
86	47.8	47.8	47.9	47.9	48.0	48.1	48.1	48.2	48.2	48.3
87	48.3	48.4	48.4	48.5	48.6	48.6	48.7	48.7	48.8	48.8
88	48.9	48.9	49.0	49.1	49.1	49.2	49.2	49.3	49.3	49.4
89	49.4	49.5	49.6	49.6	49.7	49.7	49.8	49.8	49.9	49.9

Example: Difference or Range in Temperature of 70°F = 38.9°C

Table 104 DIFFERENCES: DEGREES CELSIUS (CENTIGRADE) TO DEGREES FAHRENHEIT

[°C to °F]   1°C = 1.8°F (exactly)

°C	0.0	0.1	0.2	0.3	0.4	0.5	0.6	0.7	0.8	0.9
0	—	0.2	0.4	0.5	0.7	0.9	1.1	1.3	1.4	1.6
1	1.8	2.0	2.2	2.3	2.5	2.7	2.9	3.1	3.2	3.4
2	3.6	3.8	4.0	4.1	4.3	4.5	4.7	4.9	5.0	5.2
3	5.4	5.6	5.8	5.9	6.1	6.3	6.5	6.7	6.8	7.0
4	7.2	7.4	7.6	7.7	7.9	8.1	8.3	8.5	8.6	8.8
5	9.0	9.2	9.4	9.5	9.7	9.9	10.1	10.3	10.4	10.6
6	10.8	11.0	11.2	11.3	11.5	11.7	11.9	12.1	12.2	12.4
7	12.6	12.8	13.0	13.1	13.3	13.5	13.7	13.9	14.0	14.2
8	14.4	14.6	14.8	14.9	15.1	15.3	15.5	15.7	15.8	16.0
9	16.2	16.4	16.6	16.7	16.9	17.1	17.3	17.5	17.6	17.8
10	18.0	18.2	18.4	18.5	18.7	18.9	19.1	19.3	19.4	19.6
11	19.8	20.0	20.2	20.3	20.5	20.7	20.9	21.1	21.2	21.4
12	21.6	21.8	22.0	22.1	22.3	22.5	22.7	22.9	23.0	23.2
13	23.4	23.6	23.8	23.9	24.1	24.3	24.5	24.7	24.8	25.0
14	25.2	25.4	25.6	25.7	25.9	26.1	26.3	26.5	26.6	26.8
15	27.0	27.2	27.4	27.5	27.7	27.9	28.1	28.3	28.4	28.6
16	28.8	29.0	29.2	29.3	29.5	29.7	29.9	30.1	30.2	30.4
17	30.6	30.8	31.0	31.1	31.3	31.5	31.7	31.9	32.0	32.2
18	32.4	32.6	32.8	32.9	33.1	33.3	33.5	33.7	33.8	34.0
19	34.2	34.4	34.6	34.7	34.9	35.1	35.3	35.5	35.6	35.8
20	36.0	36.2	36.4	36.5	36.7	36.9	37.1	37.3	37.4	37.6
21	37.8	38.0	38.2	38.3	38.5	38.7	38.9	39.1	39.2	39.4
22	39.6	39.8	40.0	40.1	40.3	40.5	40.7	40.9	41.0	41.2
23	41.4	41.6	41.8	41.9	42.1	42.3	42.5	42.7	42.8	43.0
24	43.2	43.4	43.6	43.7	43.9	44.1	44.3	44.5	44.6	44.8
25	45.0	45.2	45.4	45.5	45.7	45.9	46.1	46.3	46.4	46.6
26	46.8	47.0	47.2	47.3	47.5	47.7	47.9	48.1	48.2	48.4
27	48.6	48.8	49.0	49.1	49.3	49.5	49.7	49.9	50.0	50.2
28	50.4	50.6	50.8	50.9	51.1	51.3	51.5	51.7	51.8	52.0
29	52.2	52.4	52.6	52.7	52.9	53.1	53.3	53.5	53.6	53.8
30	54.0	54.2	54.4	54.5	54.7	54.9	55.1	55.3	55.4	55.6
31	55.8	56.0	56.2	56.3	56.5	56.7	56.9	57.1	57.2	57.4
32	57.6	57.8	58.0	58.1	58.3	58.5	58.7	58.9	59.0	59.2
33	59.4	59.6	59.8	59.9	60.1	60.3	60.5	60.7	60.8	61.0
34	61.2	61.4	61.6	61.7	61.9	62.1	62.3	62.5	62.6	62.8
35	63.0	63.2	63.4	63.5	63.7	63.9	64.1	64.3	64.4	64.6
36	64.8	65.0	65.2	65.3	65.5	65.7	65.9	66.1	66.2	66.4
37	66.6	66.8	67.0	67.1	67.3	67.5	67.7	67.9	68.0	68.2
38	68.4	68.6	68.8	68.9	69.1	69.3	69.5	69.7	69.8	70.0
39	70.2	70.4	70.6	70.7	70.9	71.1	71.3	71.5	71.6	71.8
40	72.0	72.2	72.4	72.5	72.7	72.9	73.1	73.3	73.4	73.6
41	73.8	74.0	74.2	74.3	74.5	74.7	74.9	75.1	75.2	75.4
42	75.6	75.8	76.0	76.1	76.3	76.5	76.7	76.9	77.0	77.2
43	77.4	77.6	77.8	77.9	78.1	78.3	78.5	78.7	78.8	79.0
44	79.2	79.4	79.6	79.7	79.9	80.1	80.3	80.5	80.6	80.8

Example: Difference or Range in Temperature of 10.3°C = 18.5°F

Table 105 BRITISH THERMAL UNITS TO KILOJOULES
[Btu to kJ]   1 Btu = 1.055 055 853 kJ

(a)

Btu	0	100	200	300	400	500	600	700	800	900
0	—	105.51	211.01	316.52	422.02	527.53	633.03	738.54	844.04	949.55
1000	1055.06	1160.56	1266.07	1371.57	1477.08	1582.58	1688.09	1793.59	1899.10	2004.61
2000	2110.11	2215.62	2321.12	2426.63	2532.13	2637.64	2743.15	2848.65	2954.16	3059.66
3000	3165.17	3270.67	3376.18	3481.68	3587.19	3692.70	3798.20	3903.71	4009.21	4114.72
4000	4220.22	4325.73	4431.23	4536.74	4642.25	4747.75	4853.26	4958.76	5064.27	5169.77
5000	5275.28	5380.78	5486.29	5591.80	5697.30	5802.81	5908.31	6013.82	6119.32	6224.83
6000	6330.34	6435.84	6541.35	6646.85	6752.36	6857.86	6963.37	7068.87	7174.38	7279.89
7000	7385.39	7490.90	7596.40	7701.91	7807.41	7912.92	8018.42	8123.93	8229.44	8334.94
8000	8440.45	8545.95	8651.46	8756.96	8862.47	8967.97	9073.48	9178.99	9284.49	9390.00
9000	9495.50	9601.01	9706.51	9812.02	9917.53	10023.03	10128.54	10234.04	10339.55	10445.05

(b)

Btu	0	10	20	30	40	50	60	70	80	90
0	0	10.55	21.10	31.65	42.20	52.75	63.30	73.85	84.40	94.96

Example: 2650 Btu = 2743.15 + 52.75 kJ = 2795.90 kJ

Table 106 KILOJOULES TO BRITISH THERMAL UNITS
[kJ to Btu]   1 kJ = 0.947 817 12 Btu

(a)

kJ	0	100	200	300	400	500	600	700	800	900
0	—	94.78	189.56	284.35	379.13	473.91	568.69	663.47	758.25	853.04
1000	947.82	1042.60	1137.38	1232.16	1326.94	1421.73	1516.51	1611.29	1706.07	1800.85
2000	1895.63	1990.42	2085.20	2179.98	2274.76	2369.54	2464.32	2559.11	2653.89	2748.67
3000	2843.45	2938.23	3033.01	3127.80	3222.58	3317.36	3412.14	3506.92	3601.71	3696.49
4000	3791.27	3886.05	3980.83	4075.61	4170.40	4265.18	4359.96	4454.74	4549.52	4644.30
5000	4739.09	4833.87	4928.65	5023.43	5118.21	5212.99	5307.78	5402.56	5497.34	5592.12
6000	5686.90	5781.68	5876.47	5971.25	6066.03	6160.81	6255.59	6350.37	6445.16	6539.94
7000	6634.72	6729.50	6824.28	6919.06	7013.85	7108.63	7203.41	7298.19	7392.97	7487.76
8000	7582.54	7677.32	7772.10	7866.88	7961.66	8056.45	8151.23	8246.01	8340.79	8435.57
9000	8530.35	8625.14	8719.92	8814.70	8909.48	9004.26	9099.04	9193.83	9288.61	9383.39

(b)

kJ	0	10	20	30	40	50	60	70	80	90
0	—	9.48	18.96	28.43	37.91	47.39	56.87	66.35	75.83	85.30

Example: 1340 kJ = 1232.16 + 37.91 Btu = 1270.07 Btu
Note: 100 000 Btu = 1 therm

Table 107 CALORIES TO JOULES

[cal to J]   1 cal = 4.186 8 J (exactly)

(a)

cal	0	100	200	300	400	500	600	700	800	900
0	—	418.7	837.4	1256.0	1674.7	2093.4	2512.1	2930.8	3349.4	3768.1
1000	4186.8	4605.5	5024.2	5442.8	5861.5	6280.2	6698.9	7117.6	7536.2	7954.9
2000	8373.6	8792.3	9211.0	9629.6	10048.3	10467.0	10885.7	11304.4	11723.0	12141.7
3000	12560.4	12979.1	13397.8	13816.4	14235.1	14653.8	15072.5	15491.2	15909.8	16328.5
4000	16747.2	17165.9	17584.6	18003.2	18421.9	18840.6	19259.3	19678.0	20096.6	20515.3
5000	20934.0	21352.7	21771.4	22190.0	22608.7	23027.4	23446.1	23864.8	24283.4	24702.1
6000	25120.8	25539.5	25958.2	26376.8	26795.5	27214.2	27632.9	28051.6	28470.2	28888.9
7000	29307.6	29726.3	30145.0	30563.6	30982.3	31401.0	31819.7	32238.4	32657.0	33075.7
8000	33494.4	33913.1	34331.8	34750.4	35169.1	35587.8	36006.5	36425.2	36843.8	37262.5
9000	37681.2	38099.9	38518.6	38937.2	39355.9	39774.6	40193.3	40612.0	41030.6	41449.3

(b)

cal	0	10	20	30	40	50	60	70	80	90
0	—	41.9	83.7	125.6	167.5	209.3	251.2	293.1	334.9	376.8

Example: 6750 cal = 28051.6 + 209.3 J = 28260.9 J or 6.75 kcal = 28.2609 kJ
Note: The calorie used in Tables 105, 106, 107 and 108 is the international table calorie

Table 108 JOULES TO CALORIES

[J to cal]   1 J = 0.238 845 896 6 cal

(a)

J	0	10000	20000	30000	40000	50000	60000	70000	80000	90000
0	—	2388.459	4776.918	7165.377	9553.836	11942.295	14330.754	16719.213	19107.672	21496.131

(b)

J	0	100	200	300	400	500	600	700	800	900
0	—	23.885	47.769	71.654	95.538	119.423	143.308	167.192	191.077	214.961
1000	238.846	262.730	286.615	310.500	334.384	358.269	382.153	406.038	429.923	453.807
2000	477.692	501.576	525.461	549.346	573.230	597.115	620.999	644.884	668.769	692.653
3000	716.538	740.422	764.307	788.191	812.076	835.961	859.845	883.730	907.614	931.499
4000	955.384	979.268	1003.153	1027.037	1050.922	1074.807	1098.691	1122.576	1146.460	1170.345
5000	1194.229	1218.114	1241.999	1265.883	1289.768	1313.652	1337.537	1361.422	1385.306	1409.191
6000	1433.075	1456.960	1480.845	1504.729	1528.614	1552.498	1576.383	1600.268	1624.152	1648.037
7000	1671.921	1695.806	1719.690	1743.575	1767.460	1791.344	1815.229	1839.113	1862.998	1886.883
8000	1910.767	1934.652	1958.536	1982.421	2006.306	2030.190	2054.075	2077.959	2101.844	2125.728
9000	2149.613	2173.498	2197.382	2221.267	2245.151	2269.036	2292.921	2316.805	2340.690	2364.574

(c)

J	0	10	20	30	40	50	60	70	80	90
0	—	2.388	4.777	7.165	9.554	11.942	14.331	16.719	19.108	21.496

Example: 38410 J = 7165.377 + 2006.306 + 2.388 cal = 9174.071 cal or 38.41 kJ = 9.174071 kcal
Note: Table 107 can also be used for converting:
(a) kilocalories to kilojoules, and
(b) calories per cubic metre to joules per cubic metre;
and Table 108 for converting:
(a) kilojoules to kilocalories, and
(b) joules per cubic metre to calories per cubic metre
Similarly, as 1 Btu lb^{-1} degF = 1 kcal kg^{-1} degC (exactly), Table 107 can be used for converting British thermal units per pound degree F to kilojoules per kilogramme degree C, and Table 108 for vice versa

Table 109 BRITISH THERMAL UNITS PER HOUR TO KILOWATTS

[Btu h^{-1} to kW] 1 Btu h^{-1} = 0.293 071 070 2 W
= 0.000 293 071 070 2 kW

(a)

Btu h^{-1}	0	10000	20000	30000	40000	50000	60000	70000	80000	90000
0	—	2.931	5.861	8.792	11.723	14.654	17.584	20.515	23.446	26.376
100000	29.307	32.238	35.169	38.099	41.030	43.961	46.891	49.822	52.753	55.684
200000	58.614	61.545	64.476	67.406	70.337	73.268	76.198	79.129	82.060	84.991
300000	87.921	90.852	93.783	96.713	99.644	102.575	105.506	108.436	111.367	114.298
400000	117.228	120.159	123.090	126.021	128.951	131.882	134.813	137.743	140.674	143.605
500000	146.536	149.466	152.397	155.328	158.258	161.189	164.120	167.051	169.981	172.912
600000	175.843	178.773	181.704	184.635	187.565	190.496	193.427	196.358	199.288	202.219
700000	205.150	208.080	211.011	213.942	216.873	219.803	222.734	225.665	228.595	231.526
800000	234.457	237.388	240.318	243.249	246.180	249.110	252.041	254.972	257.903	260.833
900000	263.764	266.695	269.625	272.556	275.487	278.418	281.348	284.279	287.210	290.140

(b)

Btu h^{-1}	0	1000	2000	3000	4000	5000	6000	7000	8000	9000
0	—	0.293	0.586	0.879	1.172	1.465	1.758	2.051	2.345	2.638

Example: 115000 Btu h^{-1} = 32.238 + 1.465 kW = 33.703 kW

Table 110 KILOWATTS TO BRITISH THERMAL UNITS PER HOUR

[kW to Btu h^{-1}] 1 kW = 3 412.141 63 Btu h^{-1}

(a)

kW	0	10	20	30	40	50	60	70	80	90
0	—	34121	68243	102364	136486	170607	204728	238850	272971	307093
100	341214	375336	409457	443578	477700	511821	545943	580064	614185	648307
200	682428	716550	750671	784793	818914	853035	887157	921278	955400	989521
300	1023642	1057764	1091885	1126007	1160128	1194250	1228371	1262492	1296614	1330735
400	1364857	1398978	1433099	1467221	1501342	1535464	1569585	1603707	1637828	1671949
500	1706071	1740192	1774314	1808435	1842556	1876678	1910799	1944921	1979042	2013164
600	2047285	2081406	2115528	2149649	2183771	2217892	2252013	2286135	2320256	2354378
700	2388499	2422621	2456742	2490863	2524985	2559106	2593228	2627349	2661470	2695592
800	2729713	2763835	2797956	2832078	2866199	2900320	2934442	2968563	3002685	3036806
900	3070927	3105049	3139170	3173292	3207413	3241535	3275656	3309777	3343899	3378020

(b)

kW	0	1	2	3	4	5	6	7	8	9
0	—	3412	6824	10236	13649	17061	20473	23885	27297	30709

Example: 35 kW = 102364 + 17061 Btu h^{-1} = 119425 Btu h^{-1}

Table 111 BRITISH THERMAL UNITS PER POUND TO KILOJOULES PER KILOGRAM

[Btu lb^{-1} to kJ kg^{-1}]   1 Btu lb^{-1} = 2.326 kJ kg^{-1} (exactly)
All values in Tables (a) and (b) are exact

(a)

Btu lb^{-1}	0	10	20	30	40	50	60	70	80	90
0	—	23.26	46.52	69.78	93.04	116.30	139.56	162.82	186.08	209.34
100	232.60	255.86	279.12	302.38	325.64	348.90	372.16	395.42	418.68	441.94
200	465.20	488.46	511.72	534.98	558.24	581.50	604.76	628.02	651.28	674.54
300	697.80	721.06	744.32	767.58	790.84	814.10	837.36	860.62	883.88	907.14
400	930.40	953.66	976.92	1000.18	1023.44	1046.70	1069.96	1093.22	1116.48	1139.74
500	1163.00	1186.26	1209.52	1232.78	1256.04	1279.30	1302.56	1325.82	1349.08	1372.34
600	1395.60	1418.86	1442.12	1465.38	1488.64	1511.90	1535.16	1558.42	1581.68	1604.94
700	1628.20	1651.46	1674.72	1697.98	1721.24	1744.50	1767.76	1791.02	1814.28	1837.54
800	1860.80	1884.06	1907.32	1930.58	1953.84	1977.10	2000.36	2023.62	2046.88	2070.14
900	2093.40	2116.66	2139.92	2163.18	2186.44	2209.70	2232.96	2256.22	2279.48	2302.74

(b)

Btu lb^{-1}	0	1	2	3	4	5	6	7	8	9
0	—	2.326	4.652	6.978	9.304	11.630	13.956	16.282	18.608	20.934

Example: 891 Btu lb^{-1} = 2070.14 + 2.326 kJ kg^{-1} = 2072.466 kJ kg^{-1}

Table 112 KILOJOULES PER KILOGRAM TO BRITISH THERMAL UNITS PER POUND

[kJ kg^{-1} to Btu lb^{-1}]   1 kJ kg^{-1} = 0.429 922 61 Btu lb^{-1}

(a)

kJ kg^{-1}	0	100	200	300	400	500	600	700	800	900
0	—	42.992	85.985	128.977	171.969	214.961	257.954	300.946	343.938	386.930
1000	429.923	472.915	515.907	558.899	601.892	644.884	687.876	730.868	773.861	816.853
2000	859.845	902.837	945.830	988.822	1031.814	1074.807	1117.799	1160.791	1203.783	1246.776
3000	1289.768	1332.760	1375.752	1418.745	1461.737	1504.729	1547.721	1590.714	1633.706	1676.698
4000	1719.690	1762.683	1805.675	1848.667	1891.659	1934.652	1977.644	2020.636	2063.629	2106.621
5000	2149.613	2192.605	2235.598	2278.590	2321.582	2364.574	2407.567	2450.559	2493.551	2536.543
6000	2579.536	2622.528	2665.520	2708.512	2751.505	2794.497	2837.489	2880.481	2923.474	2966.466
7000	3009.458	3052.451	3095.443	3138.435	3181.427	3224.420	3267.412	3310.404	3353.396	3396.389
8000	3439.381	3482.373	3525.365	3568.358	3611.350	3654.342	3697.334	3740.327	3783.319	3826.311
9000	3869.303	3912.296	3955.288	3998.280	4041.273	4084.265	4127.257	4170.249	4213.242	4256.234

(b)

kJ kg^{-1}	0	10	20	30	40	50	60	70	80	90
0	—	4.299	8.598	12.898	17.197	21.496	25.795	30.095	34.394	38.693

Example: 3450 kJ kg^{-1} = 1461.737 + 21.496 Btu lb^{-1} = 1483.233 Btu lb^{-1}

Table 113 BRITISH THERMAL UNITS PER CUBIC FOOT TO KILOJOULES PER CUBIC METRE

[Btu ft^{-3} to kJ m^{-3}]   1 Btu ft^{-3} = 37.258 945 8 kJ m^{-3}

(a)

Btu ft^{-3}	0	10	20	30	40	50	60	70	80	90
0	—	372.59	745.18	1117.77	1490.36	1862.95	2235.54	2608.13	2980.72	3353.31
100	3725.89	4098.48	4471.07	4843.66	5216.25	5588.84	5961.43	6334.02	6706.61	7079.20
200	7451.79	7824.38	8196.97	8569.56	8942.15	9314.74	9687.33	10059.92	10432.50	10805.09
300	11177.68	11550.27	11922.86	12295.45	12668.04	13040.63	13413.22	13785.81	14158.40	14530.99
400	14903.58	15276.17	15648.76	16021.35	16393.94	16766.53	17139.12	17511.70	17884.29	18256.88
500	18629.47	19002.06	19374.65	19747.24	20119.83	20492.42	20865.01	21237.60	21610.19	21982.78
600	22355.37	22727.96	23100.55	23473.14	23845.73	24218.31	24590.90	24963.49	25336.08	25708.67
700	26081.26	26453.85	26826.44	27199.03	27571.62	27944.21	28316.80	28689.39	29061.98	29434.57
800	29807.16	30179.75	30552.34	30924.93	31297.51	31670.10	32042.69	32415.28	32787.87	33160.46
900	33533.05	33905.64	34278.23	34650.82	35023.41	35396.00	35768.59	36141.18	36513.77	36886.36

(b)

Btu ft^{-3}	0	1	2	3	4	5	6	7	8	9
0	—	37.26	74.52	111.78	149.04	186.29	223.55	260.81	298.07	335.33

Example: 555 Btu ft^{-3} = 20492.42 + 186.29 kJ m^{-3} = 20678.71 kJ m^{-3}

Table 114 KILOJOULES PER CUBIC METRE TO BRITISH THERMAL UNITS PER CUBIC FOOT

[kJ m^{-3} to Btu ft^{-3}]   1 kJ m^{-3} = 0.026 839 192 Btu ft^{-3}

(a)

kJ m^{-3}	0	100	200	300	400	500	600	700	800	900
0	—	2.6839	5.3678	8.0518	10.7357	13.4196	16.1035	18.7874	21.4714	24.1553
1000	26.8392	29.5231	32.2070	34.8909	37.5749	40.2588	42.9427	45.6266	48.3105	50.9945
2000	53.6784	56.3623	59.0462	61.7301	64.4141	67.0980	69.7819	72.4658	75.1497	77.8337
3000	80.5176	83.2015	85.8854	88.5693	91.2533	93.9372	96.6211	99.3050	101.9889	104.6728
4000	107.3568	110.0407	112.7246	115.4085	118.0924	120.7764	123.4603	126.1442	128.8281	131.5120
5000	134.1960	136.8799	139.5638	142.2477	144.9316	147.6156	150.2995	152.9834	155.6673	158.3512
6000	161.0352	163.7191	166.4030	169.0869	171.7708	174.4547	177.1387	179.8226	182.5065	185.1904
7000	187.8743	190.5583	193.2422	195.9261	198.6100	201.2939	203.9779	206.6618	209.3457	212.0296
8000	214.7135	217.3975	220.0814	222.7653	225.4492	228.1331	230.8171	233.5010	236.1849	238.8688
9000	241.5527	244.2366	246.9206	249.6045	252.2884	254.9723	257.6562	260.3402	263.0241	265.7080

(b)

kJ m^{-3}	0	10	20	30	40	50	60	70	80	90
0	—	0.2684	0.5368	0.8052	1.0736	1.3420	1.6104	1.8787	2.1471	2.4155

Example: 2440 kJ m^{-3} = 64.4141 + 1.0736 Btu ft^{-3} = 65.4877 Btu ft^{-3}

Table 115 BRITISH THERMAL UNITS PER SQUARE FOOT HOUR TO KILOJOULES PER SQUARE METRE HOUR

[Btu ft^{-2} h^{-1} to kJ m^{-2} h^{-1}] 1 Btu ft^{-2} h^{-1} = 11.356 527 kJ m^{-2} h^{-1}

(a)

Btu ft^{-2}h^{-1}	0	10	20	30	40	50	60	70	80	90
0	—	113.57	227.13	340.70	454.26	567.83	681.39	794.96	908.52	1022.09
100	1135.65	1249.22	1362.78	1476.35	1589.91	1703.48	1817.04	1930.61	2044.17	2157.74
200	2271.31	2384.87	2498.44	2612.00	2725.57	2839.13	2952.70	3066.26	3179.83	3293.39
300	3406.96	3520.52	3634.09	3747.65	3861.22	3974.78	4088.35	4201.91	4315.48	4429.05
400	4542.61	4656.18	4769.74	4883.31	4996.87	5110.44	5224.00	5337.57	5451.13	5564.70
500	5678.26	5791.83	5905.39	6018.96	6132.52	6246.09	6359.66	6473.22	6586.79	6700.35
600	6813.92	6927.48	7041.05	7154.61	7268.18	7381.74	7495.31	7608.87	7722.44	7836.00
700	7949.57	8063.13	8176.70	8290.26	8403.83	8517.40	8630.96	8744.53	8858.09	8971.66
800	9085.22	9198.79	9312.35	9425.92	9539.48	9653.05	9766.61	9880.18	9993.74	10107.31
900	10220.87	10334.44	10448.00	10561.57	10675.14	10788.70	10902.27	11015.83	11129.40	11242.96

(b)

Btu ft^{-2}h^{-1}	0	1	2	3	4	5	6	7	8	9
0	—	11.36	22.71	34.07	45.43	56.78	68.14	79.50	90.85	102.21

Example: 185 Btu ft^{-2} h^{-1} = 2044.17 + 56.78 kJ m^{-2} h^{-1} = 2100.95 kJ m^{-2} h^{-1}

Table 116 KILOJOULES PER SQUARE METRE HOUR TO BRITISH THERMAL UNITS PER SQUARE FOOT HOUR

[kJ m^{-2} h^{-1} to Btu ft^{-2} h^{-1}] 1 kJ m^{-2} h^{-1} = 0.088 055 09 Btu ft^{-2} h^{-1}

(a)

kJ m^{-2}h^{-1}	0	100	200	300	400	500	600	700	800	900
0	—	8.806	17.611	26.417	35.222	44.028	52.833	61.639	70.444	79.250
1000	88.055	96.861	105.666	114.472	123.277	132.083	140.888	149.694	158.499	167.305
2000	176.110	184.916	193.721	202.527	211.332	220.138	228.943	237.749	246.554	255.360
3000	264.165	272.971	281.776	290.582	299.387	308.193	316.998	325.804	334.609	343.415
4000	352.220	361.026	369.831	378.637	387.442	396.248	405.053	413.859	422.664	431.470
5000	440.275	449.081	457.886	466.692	475.497	484.303	493.109	501.914	510.720	519.525
6000	528.331	537.136	545.942	554.747	563.553	572.358	581.164	589.969	598.775	607.580
7000	616.386	625.191	633.997	642.802	651.608	660.413	669.219	678.024	686.830	695.635
8000	704.441	713.246	722.052	730.857	739.663	748.468	757.274	766.079	774.885	783.690
9000	792.496	801.301	810.107	818.912	827.718	836.523	845.329	854.134	862.940	871.745

(b)

kJ m^{-2} h^{-1}	0	10	20	30	40	50	60	70	80	90
0	—	0.881	1.761	2.642	3.522	4.403	5.283	6.164	7.044	7.925

Example: 4260 kJ m^{-2} h^{-1} = 369.831 + 5.283 Btu ft^{-2} h^{-1} = 375.114 Btu ft^{-2} h^{-1}.

11